LONDON MATHEMATICAL SOCIETY LECTURE NOTE SERIES

Editor: PROFESSOR G. C. SHEPHARD, University of East Anglia

This series publishes the records of lectures and seminars on advanced topics in mathematics held at universities throughout the world. For the most part, these are at postgraduate level either presenting new material or describing older material in a new way. Exceptionally, topics at the undergraduate level may be published if the treatment is sufficiently original.

Prospective authors should contact the editor in the first instance.

Already published in this series

1. General cohomology theory and K-theory, PETER HILTON.
2. Numerical ranges of operators on normed spaces and of elements of normed algebras, F. F. BONSALL and J. DUNCAN.
3. Convex polytopes and the upper bound conjecture, P. McMULLEN and G. C. SHEPHARD.
4. Algebraic topology: A student's guide, J. F. ADAMS.
5. Commutative algebra, J. T. KNIGHT.
6. Finite groups of automorphisms, NORMAN BIGGS.
7. Introduction to combinatory logic, J. R. HINDLEY, B. LERCHER and J. P. SELDIN.
8. Integration and harmonic analysis on compact groups, R. E. EDWARDS.
9. Elliptic functions and elliptic curves, PATRICK DU VAL.
10. Numerical ranges II, F. F. BONSALL and J. DUNCAN.
11. New developments in topology, G. SEGAL (ed.).
12. Symposium on complex analysis Canterbury 1973, J. CLUNIE and W. K. HAYMAN (eds.).
13. Combinatorics, Proceedings of the British combinatorial conference 1973, T. P. McDONOUGH and V. C. MAVRON (eds.).
14. Analytic theory of abelian varieties, H. P. F. SWINNERTON-DYER.
15. An introduction to topological groups, P. J. HIGGINS.
16. Topics in finite groups, TERENCE M. GAGEN.
17. Differentiable germs and catastrophes, THEODOR BRÖCKER and L. LANDER.
18. A geometric approach to homology theory, S. BUONCRISTIANO, C. P. ROURKE and B. J. SANDERSON.
19. Graph theory, coding theory and block designs, P. J. CAMERON and J. H. VAN LINT.
20. Sheaf theory, B. R. TENNISON.
21. Automatic continuity, A. M. SINCLAIR.
22. Presentations of groups, D. L. JOHNSON.
23. Parallelisms of complete designs, PETER J. CAMERON.

LONDON MATHEMATICAL SOCIETY LECTURE NOTE SERIES

[illegible]

London Mathematical Society Lecture Note Series. 24

The Topology of Stiefel Manifolds

I.M. JAMES

Savilian Professor of Geometry
Mathematical Institute
University of Oxford

CAMBRIDGE UNIVERSITY PRESS

CAMBRIDGE

LONDON NEW YORK MELBOURNE

CAMBRIDGE UNIVERSITY PRESS
Cambridge, New York, Melbourne, Madrid, Cape Town, Singapore, São Paulo

Cambridge University Press
The Edinburgh Building, Cambridge CB2 8RU, UK

Published in the United States of America by Cambridge University Press, New York

www.cambridge.org
Information on this title: www.cambridge.org/9780521213349

First published 1976
Re-issued in this digitally printed version 2007

A catalogue record for this publication is available from the British Library

Library of Congress Catalogue Card Number: 76-9546

ISBN 978-0-521-21334-9 paperback

Contents

		page
	Preface	vii
1.	Introduction: algebra versus topology	1
2.	The Stiefel manifolds	13
3.	The auxiliary spaces	21
4.	Retractible fibrations	27
5.	Thom spaces	33
6.	Homotopy equivariance	40
7.	Cross-sections and the S-type	45
8.	Relative Stiefel manifolds	52
9.	Cannibalistic characteristic classes	57
10.	Exponential characteristic classes	61
11.	The main theorem of J-theory	71
12.	The fibre suspension	78
13.	Canonical automorphisms	83
14.	The iterated suspension	89
15.	Samelson products	94
16.	The Hopf construction	100
17.	The Bott suspension	109
18.	The intrinsic join again	116
19.	Homotopy-commutativity	123
20.	The triviality problem	129
21.	When is $P_{n,k}$ neutral?	133
22.	When is $V_{n,2}$ neutral?	139
23.	When is $V_{n,k}$ neutral?	143
24.	Further results and problems	150
	Bibliography	155
	Index	167

Preface

These lectures originated in a course given at Harvard in 1961. Algebraic topology has advanced a long way since that time. Throughout mathematics, the right kind of problem provides the challenge which leads to the improvement of technique and the development of new methods. To a considerable extent, problems about Stiefel manifolds have performed this function in algebraic topology. Thus I felt it might be useful to bring my lectures up-to-date and give some account of what is now known.

The basic theory necessary can be found in a number of text books, such as that of Spanier [132]. At appropriate places I have summarized such additional theory as is needed, with references to the literature, in the hope that these notes may be accessible to non-specialists and particularly to graduate students. Many examples are given and further problems suggested.

The literature on Stiefel manifolds is extensive, as the bibliography at the end of these notes will indicate. The topics I have chosen to discuss in detail are mainly those I have worked on myself, but as well as my own papers I have drawn on those by Adams, Atiyah, Bott and many others. Although much of the material has been published before, in some shape or form, there is a fair amount which has not. The section on further development contains information about work by Friedlander, Gitler, Mahowald, Milgram, Zvengrowski and others which is in process of publication; I am very grateful to those concerned for communicating these results. These notes were read in draft form by Sutherland, Woodward and Zvengrowski, whose comments have been most helpful. I would also like to thank Wilson Sutherland and Emery Thomas for allowing me to quote from joint work, and to thank the American Mathematical Society, Clarendon Press, London Mathematical

Society and Pergamon Press for permission to draw on previously published material.

Oxford University Mathematical Institute

Take the topological product $S^n \times S^n$ of the n-sphere with itself. Remove the diagonal and the antidiagonal. What is left is the space X_n of pairs (x, y) such that $x \neq \pm y$. For what values of n is it possible to make a continuous deformation of X_n into itself in which each such pair (x, y) is deformed into the pair (y, x)? It is known that the deformation is impossible unless $n + 1$ is a power of two; and that the deformation is possible for $n = 1, 3, 7, 15$ and 31; the position for $n = 63, 127, \ldots$ is at present unknown.

1·Introduction: algebra versus topology

There are three families of Stiefel manifolds, the real, the complex and the quaternionic. Readers of these notes may already be familiar with the account of their basic properties to be found in standard texts such as Steenrod [133] and Steenrod-Epstein [134]; a summary is given in §2 below. In this introduction we shall only be dealing with the real family, which is undoubtedly the most interesting. Some of the real Stiefel manifolds have particular topological properties, due to the existence of certain constructions which are algebraic in origin. Our aim is to try and understand, from the topological point of view, why some of them have these properties while others do not.

The notation we use is fairly standard. Thus R^m denotes euclidean m-space $(m = 0, 1, \ldots)$ with the usual embedding of R^m in R^{m+1}. The vectors $v \in R^m$ such that $|v| \leq 1$ form the unit ball B^m and those such that $|v| = 1$ form the unit sphere S^{m-1}. The projective space P^{m-1} is obtained from S^{m-1} by identifying v with $-v$ for all $v \in S^{m-1}$. The group of orthogonal transformations of R^m is denoted by O_m. Thus $P^{m-1} \subset P^m$ and $O_m \subset O_{m+1}$, in the usual way. Unless it is necessary to be more specific the basepoint in any space is denoted by e; orientation conventions are as in [64], and $\iota_m \in \pi_m(S^m)$ denotes the class of the identity self-map.

Following Stiefel [136] and many others let $V_{n,k}$, where $1 \leq k \leq n$, denote the manifold of orthonormal k-frames in R^n. Elements of $V_{n,k}$ correspond, in an obvious way, to norm-preserving linear transformations of R^k into R^n. The orthogonal group O_k acts on $V_{n,k}$ by pre-composition, while the orthogonal group O_n acts on $V_{n,k}$ by post-composition. The latter action is transitive and enables $V_{n,k}$ to be identified with the factor space of O_n by O_{n-k}. For $k < n$ the rotation group can be used instead of the full orthogonal group.

If we pre- or post-compose with a rotation we obtain a self-map of $V_{n,k}$ in the homotopy class of the identity. If we pre-compose by a non-rotation we obtain a self-map of homotopy class λ, say; if we post-compose by a non-rotation we obtain a self-map of homotopy class μ, say. In the semigroup of homotopy classes of self-maps of $V_{n,k}$ these canonical classes satisfy the relations

$$(1.1)\quad \lambda^2 = 1 = \mu^2, \quad \lambda\mu = \mu\lambda, \quad \lambda^k = \mu^n.$$

To prove the last of these, represent elements of $V_{n,k}$ by matrices with k columns - the vectors of the k-frame - and n rows. The class λ includes the self-maps which change the sign of any column. The class μ includes the self-maps which change the sign of any row. Since changing the sign of all the columns has the same effect as changing the sign of all the rows we obtain $\lambda^k = \mu^n$, as asserted. Note that $\lambda = 1$ if n is even and k odd, while $\mu = 1$ if n is odd and k even. In some applications it is the class $\xi = \lambda\mu$ which is important; note that $\xi = 1$ if n and k are both odd.

We can fibre $V_{n,k}$ over $V_{n,1} = S^{n-1}$ by taking one vector (say the last) from each k-frame. A cross-section $f : S^{n-1} \to V_{n,k}$ associates with each point $v \in S^{n-1}$ an orthonormal k-frame $(v_1, \ldots, v_{k-1}, v)$. Thus $(v_1, \ldots, v_{k-1}) = g(v)$, say, is an orthonormal $(k - 1)$-frame; we refer to $g : S^{n-1} \to V_{n,k-1}$ as the projection of f. We can always regard $(v_1, \ldots, v_{k-1})$ as a $(k - 1)$-frame of tangents to S^{n-1} at the point v. Hence a cross-section of $V_{n,k}$ over S^{n-1} is equivalent to an (orthonormal) $(k - 1)$-field on S^{n-1}, i.e. a field of orthonormal tangent $(k - 1)$-frames. Any such $(k - 1)$-field spans a field of tangent $(k - 1)$-planes. Conversely Steenrod has shown, in §27 of [133], that if S^{n-1} admits a field of tangent $(k - 1)$-planes and $2k \leq n + 1$ then S^{n-1} admits a $(k - 1)$-field. This does not mean, however, that every field of tangent $(k - 1)$-planes can be spanned by a $(k - 1)$-field (see [62]).

For what values of n and k does $V_{n,k}$ admit a cross-section, over S^{n-1}? Take $k = 2$, for example. We need to find a self-map g of S^{n-1} such that $g(v)$ is orthogonal to v, for all $v \in S^{n-1}$. When n is even, say $n = 2m$, we can regard v as a complex m-vector, rather than a real $2m$-vector, and define g through multiplication by i. In terms of coordinates, if $v = (x_0, x_1, \ldots, x_{2m-2}, x_{2m-1})$ then

$g(v) = (-x_1, x_0, \ldots, -x_{2m-1}, x_{2m-2})$. Conversely, suppose that g exists, with $g(v)$ orthogonal to v. Then

$$h_t(v) = v \cos \pi t + g(v) \sin \pi t \qquad (0 \le t \le 1)$$

defines a homotopy between the identity on S^{n-1} and the antipodal map. Since the degree of the latter is $(-1)^n$ it follows at once that n is even. Thus $V_{n,2}$ admits a cross-section if and only if n is even.

Given k, we can construct cross-sections of $V_{n,k}$ for suitable values of n as follows. Consider the Clifford algebra C_m $(m = 0, 1, \ldots)$ generated by an anticommuting set of elements $(e_1, \ldots, e_m)$ such that

$$e_1^2 = \ldots = e_m^2 = -1.$$

Thus $C_0 = R$, the real numbers; $C_1 = C$, the complex numbers; and $C_2 = H$, the quaternions. The next five Clifford algebras are easily shown to be

$$H \oplus H,\ H(2),\ C(4),\ R(8),\ R(8) \oplus R(8),$$

where $A(q)$, for any algebra A and positive integer q, denotes the qth order matrix algebra over A. Moreover (see [9], for example) the matrix algebra $C_m(16)$ of order 16 over C_m is isomorphic to C_{m+8}. Thus all the Clifford algebras can be expressed in terms of matrix algebras over R, C or H.

Let $\sigma(k)$ denote the number of integers s in the range $0 < s < k$ such that $s \equiv 0, 1, 2$ or $4 \bmod 8$. Clearly R^n can be represented as a C_{k-1}-module whenever $n \equiv 0 \bmod a_k$, where $a_k = 2^{\sigma(k)}$. Any such representation can be orthogonalized, in the usual way, so that the generators $e_1, \ldots, e_{k-1}$ correspond to orthogonal transformations, and then a cross-section $f : S^{n-1} \to V_{n,k}$ is given by

$$f(v) = (e_1 . v, \ldots, e_{k-1} . v, v) \qquad (v \in S^{n-1}).$$

The existence of these Clifford cross-sections was noted by Eckmann [38], with reference to the algebraic results of Hurwitz [60] and Radon [118]. We give an example, due to Zvengrowski, of a Clifford cross-section of

$V_{16,9}$ (the first eight column vectors are tangent to S^{15} at the points given by the last).

$$\begin{array}{ccccccccc}
x_8 & -x_7 & -x_6 & -x_5 & -x_4 & -x_3 & -x_2 & -x_1 & x_0 \\
-x_9 & x_6 & -x_7 & -x_4 & x_5 & -x_2 & x_3 & x_0 & x_1 \\
-x_{10} & -x_5 & -x_4 & x_7 & x_6 & x_1 & x_0 & -x_3 & x_2 \\
-x_{11} & -x_4 & x_5 & -x_6 & x_7 & x_0 & -x_1 & x_2 & x_3 \\
-x_{12} & x_3 & x_2 & x_1 & x_0 & -x_7 & -x_6 & -x_5 & x_4 \\
-x_{13} & x_2 & -x_3 & x_0 & -x_1 & x_6 & -x_7 & x_4 & x_5 \\
-x_{14} & -x_1 & x_0 & x_3 & -x_2 & -x_5 & x_6 & x_7 & x_6 \\
-x_{15} & x_0 & x_1 & -x_2 & -x_3 & x_4 & x_5 & -x_6 & x_7 \\
-x_0 & -x_{15} & -x_{14} & -x_{13} & -x_{12} & -x_{11} & -x_{10} & -x_9 & x_8 \\
x_1 & -x_{14} & x_{15} & x_{12} & -x_{13} & x_{10} & -x_{11} & x_8 & x_9 \\
x_2 & x_{13} & x_{12} & -x_{15} & -x_{14} & -x_9 & x_8 & x_{11} & x_{10} \\
x_3 & x_{12} & -x_{13} & x_{14} & -x_{15} & x_8 & x_9 & -x_{10} & x_{11} \\
x_4 & -x_{11} & -x_{10} & -x_9 & x_8 & x_{15} & x_{14} & x_{13} & x_{12} \\
x_5 & -x_{10} & x_{11} & x_8 & x_9 & -x_{14} & x_{15} & -x_{12} & x_{13} \\
x_6 & x_9 & x_8 & -x_{11} & x_{10} & x_{13} & -x_{12} & -x_{15} & x_{14} \\
x_7 & x_8 & -x_9 & x_{10} & x_{11} & -x_{12} & -x_{13} & x_{14} & x_{15}
\end{array}$$

It was Adams [3] who finally proved the long-conjectured

Theorem (1.2). <u>The Stiefel manifold</u> $V_{n,k}$ <u>admits a cross-section, over</u> S^{n-1}, <u>if and only if</u> $n \equiv 0 \bmod a_k$.

Sufficiency we have already established. Necessity is trivial for $k = 1$ and true for $k = 2$, as we have seen. For higher values of k various results were obtained by G. W. Whitehead [153], N. E. Steenrod and J. H. C. Whitehead [135], amongst others. To indicate the kind of methods used in this subject we shall now give the proof of (1.2) in case (i) $k - 1$ is a power two or (ii) $k \not\equiv 3 \bmod 8$. In particular we prove (1.2) for all $k \leq 10$. The remaining cases are more difficult and will be dealt with later.

The Stiefel manifold $V_{n,k}$ contains a subspace $P_{n,k}$ which plays a major role in what follows. To define $P_{n,k}$, first consider the real projective $(n-1)$-space $P^{n-1} = S^{n-1}/Z_2$. Any point $\pm x \in P^{n-1}$, where $x = (x_1, \ldots, x_n)$, determines a matrix

$$\|\delta_{ij} - 2x_i x_j\| \quad (i = n - k + 1, \ldots, n; j = 1, \ldots, n).$$

The k column vectors of this matrix constitute an orthonormal k-frame in R^n, i.e. an element of $V_{n,k}$. All points of the subspace $P^{n-k-1} \subset P^{n-1}$ spanned by the first $n - k$ coordinates determine the same element of $V_{n,k}$. We define $P_{n,k}$ to be the space P^{n-1}/P^{n-k-1} obtained from P^{n-1} by collapsing P^{n-k-1} to a point and regard $P_{n,k}$ as a subspace of $V_{n,k}$ under the embedding just described. When $k = n$ we interpret $P_{n,n}$ as the space obtained from P^{n-1} by adjoining a point corresponding to the identity matrix. Notice that $P_{n,1} = V_{n,1}$. In §3 below we shall prove

Proposition (1.3). The pair $(V_{n,k}, P_{n,k})$ is $(2n - 2k)$-connected.

In fact the pair can be given CW-structure so that $V_{n,k}$ is obtained from $P_{n,k}$ by attaching cells of dimension $2n - 2k + 1$ and higher. Now let S denote the suspension functor. A simple geometric construction, as follows, enables us to prove

Proposition (1.4). If $V_{n,k}$ has a cross-section then $S^n P_{m,k}$ has the same homotopy type as $P_{m+n,k}$ for all $m \geq k$.

Let $f : S^{n-1} \to V_{n,k}$ be a cross-section and let f_v, for $v \in S^{n-1}$, denote the norm-preserving transformation $R^k \to R^n$ corresponding to $f(v)$. Consider the map

$$\theta : B^n \times R^{m-k} \times R^k \to R^{m+n-k} \times R^k$$

which is given by

$$\theta(tv, y, z) = (y, tf_v(z), (1 - t^2)^{\frac{1}{2}} z)$$

where $0 \leq t \leq 1$ and $y \in R^{m-k}$, $z \in R^k$. Since $\theta(tv, -y, -z) = -\theta(tv, y, z)$, $|\theta(tv, y, z)| = |(y, z)|$, it follows that θ induces a map

$$(B^n \times P^{m-1}, B^n \times P^{m-k-1} \cup S^{n-1} \times P^{m-1}) \to (P^{m+n-1}, P^{m+n-k-1}),$$

and hence a map

$$\phi : (B^n/S^{n-1}) \wedge (P^{m-1}/P^{m-k-1}) \to (P^{m+n-1}/P^{m+n-k-1})$$

where $\wedge$ denotes the smash product. If f is a Clifford cross-section then ϕ is a homeomorphism. In the general case it can easily be shown (see §6) that ϕ induces an isomorphism in homology and hence is a homotopy equivalence, by the theorem of J. H. C. Whitehead [159].

Let us now see what information can be extracted from (1. 4) by using the Steenrod squares in mod 2 cohomology. Recall that

$$H^*(P^{n-1}) = Z_2[a] \bmod a^n,$$

where a generates $H^1(P^{n-1})$, and that

$$Sq^i a^j = \binom{j}{i} a^{i+j},$$

by the Cartan product formula. From the cohomology exact sequence of the cofibration

$$P^{n-k-1} \to P^{n-1} \to P_{n,k}$$

we see that $\tilde{H}^r(P_{n,k})$, for $n - k \le r < n$, is generated by an element a_r, where

$$(1.5) \quad Sq^i a_j = \binom{j}{i} a_{i+j}$$

for $j \ge n - k$ and $i + j < n$. With (1. 4) in mind we prove

Proposition (1. 6). Given n and k suppose that $S^n P_{m,k}$ and $P_{m+n,k}$ have the same homotopy type for all $m > k$. If $k = 2^s + 1$, for some s, then $n \equiv 0 \bmod 2^{s+1}$.

Choose $m > k$ so that $m \equiv k \bmod 2^{s+1}$. Then $Sq^i H^{m-k}(P_{m,k}) = 0$, by (1. 5), for all $i > 0$. If n is an odd multiple of 2^r, where $r \le s$, then $Sq^i H^{m+n-k}(P_{m+n,k}) \ne 0$ for $i = 2^r$, hence $S^n P_{m,k}$ and $P_{m+n,k}$ are not of the same homotopy type, since Sq^i commutes with suspension. This contradiction establishes (1. 6) and hence, using (1. 4), proves (1. 2) when $k - 1$ is a power of two. The original argument of Steenrod and Whitehead is similar, except that (1. 3) is used instead of (1. 4).

Let us now replace cohomology by the functor $\tilde{K}_R$ formed from real vector bundles over a given space. Recall (see [9]) that $\tilde{K}_R(P^{n-1})$ is cyclic of order a_n with generator $\alpha = [L] - 1$, where L denotes the Hopf line bundle over P^{n-1}, and that $L^2 = L \otimes L$ is trivial. For any integer t the Adams operation ψ^t is defined, as in [3], and has the property that $\psi^t[L] = [L^t]$. Hence $\psi^t \alpha = 0$ or α according as t is even or odd. Just as in cohomology the exact sequence of the cofibration enables $\tilde{K}_R(P_{n,k})$ to be calculated. Provided $n \not\equiv k \bmod 4$ we find that $\tilde{K}_R(P_{n,k})$ can be identified with the subgroup of $\tilde{K}_R(P^{n-1})$ generated by $a_{n-k}\alpha$; when $n \equiv k \bmod 4$ there is an extra summand which complicates matters. Moreover $\psi^t = 0$ or 1 according as t is even or odd.

Let $\tau(k)$ denote the number of integers s in the range $0 < s < k$ such that $s \equiv 0, 1, 3$ or $5 \bmod 8$. Thus $\tau(k) = \sigma(k) - 1$ for $k \equiv \pm 3 \bmod 8$, and $\tau(k) = \sigma(k)$ otherwise. We prove

Proposition (1. 7). *Given $n \equiv 0 \bmod 8$ and k, suppose that $S^n P_{m,k}$ and $P_{m+n,k}$ have the same homotopy type for all $m > k$. Then n is divisible by $2^{\tau(k)}$.*

Choose $m > k$ so that $m \not\equiv k \bmod 4$ and write $\sigma(m) - \sigma(m-k+1) = f$. Recall that $\psi^t (S^*)^n = t^{n/2} (S^*)^n \psi^t$, for all values of t, where

$$(S^*)^n : \tilde{K}_R(P_{m,k}) \approx \tilde{K}_R(S^n P_{m,k}).$$

Let t be odd. Then $\psi^t = 1$ in the domain, as we have seen, and so $\psi^t = t^{n/2}$ in the codomain. On the other hand $\psi^t = 1$ in $\tilde{K}_R(P_{m+n,k})$. Since all these groups are cyclic of order 2^f this implies that

$$(1.\,8) \quad t^{n/2} \equiv 1 \bmod 2^f.$$

However if n is an odd multiple of 2^{e-2}, for any $e \geq 2$, then

$$(1.\,9) \quad 3^{n/2} - 1 \equiv 2^{e-1} \bmod 2^e,$$

by an elementary calculation as in §8 of [3]. Putting $t = 3$ we obtain an immediate contradiction unless n is an even multiple of 2^{f-2}. However m can be chosen, with $m \not\equiv k \bmod 4$, so that $f - 1 = 2^{\tau(k)}$, and so (1. 7) is proved.

To obtain (1.2) for $k \not\equiv \pm 3 \bmod 8$ we use (1.6), with (1.4), to deal with the cases $k \leq 4$ and to show that $n \equiv 0 \bmod 8$ when $k \geq 5$; then we use (1.7), with (1.4), to complete the proof. The original proof of (1.2) by Adams is similar, except that (1.3) and other results are used instead of (1.4).

Not every cross-section is homotopic to a Clifford cross-section, as can easily be seen, but a recent result of Milgram and Zvengrowski [111] is of interest here. A cross-section $f : S^{n-1} \to V_{n,k}$ is said to be <u>skew</u> if $f(v) = (v_1, \ldots, v_k)$ implies that $f(-v) = (-v_1, \ldots, -v_k)$. For example, Clifford cross-sections have this property. Milgram and Zvengrowski show that every cross-section is homotopic to a skew cross-section.

Another kind of cross-section is as follows. Consider the self-map T of $V_{n,k}$ which changes the sign of the last vector in each k-frame. Thus T is the antipodal map in the case of $V_{n,1} = S^{n-1}$. Let us say that a cross-section $f : S^{n-1} \to V_{n,k}$ is <u>homotopy-equivariant</u> if $Tf \simeq fT$. The case $k = 1$ is trivial. When $k \geq 2$ the condition can be taken as $Tf \simeq f$, since no cross-section exists unless n is even. If k is odd and n is even then $T \simeq 1$ on $V_{n,k}$, by (1.1). Hence the interest resides in the case k even. Notice that a cross-section of $V_{n,k+1}$ determines a homotopy-equivariant cross-section of $V_{n,k}$. For let $f_1, \ldots, f_k : S^{n-1} \to S^{n-1}$ be the first k components of a cross-section of $V_{n,k+1}$, and write $h_t(v) = (f_1 v, \ldots, f_{k-1} v, v \cos \pi t + f_k v \sin \pi t)$. Then h_0 is a cross-section of $V_{n,k}$ such that $h_0 \simeq h_1 = Th_0$. In §8 and §9 below we shall prove

Theorem (1.10). <u>There exists a homotopy-equivariant cross-section of</u> $V_{n,k}$ <u>if and only if</u> $n \equiv 0 \bmod \hat{a}_k$, <u>where</u> $\hat{a}_k = a_{k+1} = 2a_k$ <u>for</u> $k = 2$ <u>or</u> $k \equiv 0 \bmod 4$, <u>and</u> $\hat{a}_k = a_k$ <u>otherwise.</u>

Let us now turn to some problems which have not yet been solved. By general theory (see [133]) $V_{n,k}$ is trivial as a fibre bundle over S^{n-1} if and only if the associated principal bundle $V_{n,n} = O_n$ admits a cross-section, i.e. if and only if $n = 2$, 4 or 8. Thus $V_{4,k}$ $(k \leq 4)$ and $V_{8,k}$ $(k \leq 8)$ are trivial as fibre bundles. For triviality in the sense of fibre homotopy type, however, nothing is known beyond

Theorem (1.11). _If $V_{n,k}$ is trivial as a fibre space over S^{n-1} then $n = 2^r$ for some $r \geq \sigma(k)$. Furthermore if k is even then $n = 2$, 4 or 8._

The proof will be given in §20 below. It is tempting to conjecture that $V_{n,k}$ is non-trivial as a fibre space if it is non-trivial as a fibre bundle: the first unsettled case is that of $V_{16,3}$. As Scheerer [124] has pointed out the solution to this problem is important for the homotopy classification of Hopf homogeneous spaces.

Another unsolved problem concerns the self-map T of $V_{n,k}$ which changes the sign of the last vector in each k-frame. Let us say that $V_{n,k}$ is _neutral_ (elsewhere _row-simple_) if $\lambda = 1$, where λ denotes the homotopy class of T, as before. Thus $V_{n,k}$ is neutral, by (1.1), whenever n is even and k odd. Moreover $V_{n,k}$ is neutral when $n = 3$ or 7 and k is even, since then $V_{n,k}$ is a retract of $V_{n+1,k+1}$, as remarked above. In §21 below we shall prove

Theorem (1.12). _Let n be odd and k even. If $V_{n,k}$ is neutral then either $n + 1$ or $k - n + 1$ is divisible by 2^t, where t denotes the least integer such that $2^t > k$._

This gives no information when $k = 2$. However, in §22 we shall prove

Theorem (1.13). _Let n be odd. Then $V_{n,2}$ is neutral if and only if the Whitehead square $w_n \in \pi_{2n-1}(S^n)$ can be halved._

Here w_n denotes the Whitehead product of the generator $\iota_n \in \pi_n(S^n)$ with itself. This vanishes, as is well known, if and only if $n = 1$, 3 or 7. Toda [144] has shown that w_{15} can be halved and Mahowald, in unpublished work, that w_{31} can be halved. It is not difficult to show that w_n $(n > 2)$ cannot be halved unless $n + 1$ is a power of two: in §23 below we shall prove

Theorem (1.14). _Let n be odd and let $n \geq 2k - 2$, where $k = 2$, 4 or 8. If $V_{n,k}$ is neutral then $n + 1$ is a power of two._

It seems reasonable to conjecture that (1.14) is true for all even values of k.

Finally let us take another look at the problem of the existence of cross-sections. Suppose that we fibre $V_{n,k}$ over $V_{n,k-1}$ by taking the last $k - 1$ vectors of each k-frame to form a $(k - 1)$-frame. Any map $f : V_{n,k-1} \to V_{n,k}$ determines a map $g : V_{n,k-1} \to S^{n-1}$, by taking the first vector of each k-frame, and f is a cross-section if and only if the vector $g(v_1, \ldots, v_{k-1})$ is orthogonal to $v_1, \ldots, v_{k-1}$ for every orthonormal $(k - 1)$-frame $(v_1, \ldots, v_{k-1})$ in n-space. When $n = 3$ or 7 and $k = 3$ such a map g can be defined as follows. Elements of R^{n+1} can be regarded as quaternions when $n = 3$, as Cayley numbers when $n = 7$. Moreover the pure elements of the algebra (i. e. those with real part zero) determine a subspace which we identify with R^n. If u and v are pure then uv is orthogonal to both u and v; moreover uv is pure when u and v are themselves orthogonal. Hence a map g with the desired properties is defined by $g(u, v) = uv$. Thus $V_{n,3}$ admits a cross-section over $V_{n,2}$ for $n = 3$ or 7.

Conversely, if $V_{n,3}$ admits a cross-section over $V_{n,2}$ then S^n is an H-space and so $n = 3$ or 7, by the main theorem of Adams [1]. To see this, consider the unit ball $B^n \subset R^n$, of which S^{n-1} is the boundary, and the sphere S^n, of which S^{n-1} is the equator and $\pm e$, say, the poles. Given a cross-section $f : V_{n,2} \to V_{n,3}$ with projection $g : V_{n,2} \to S^{n-1}$, let $g' : B^n \times B^n \to B^n$ denote the map defined by

$$g'(au, bv) = ab \sin\theta \, g(u, (v - u\cos\theta)/\sin\theta),$$

where $u, v \in S^{n-1}$ and $a, b \in I = [0, 1]$, also $\cos\theta = u.v$, the inner product, for $0 \leq \theta \leq \pi$. Now let $h : S^n \times S^n \to S^n$ be defined by $h(\alpha e + x, \beta e + y) = (\alpha\beta e - x.y + \alpha y + \beta x + g'(x, y)$, where $x, y \in B^n$ and $-1 \leq \alpha, \beta \leq 1$. Clearly $h(\alpha e + x, e) = \alpha e + x$, $h(e, \beta e + y) = \beta e + y$, and so h constitutes an H-structure on S^n. Of course this construction is modelled on the formulae for quaternionic and Cayley multiplication contained in the previous paragraph. Summing up, we have proved

Theorem (1.15). <u>There exists a cross-section of</u> $V_{n,3}$ <u>over</u> $V_{n,2}$ <u>if and only if</u> $n = 3$ <u>or</u> 7.

Cross-sections of $V_{8,4}$ over $V_{8,3}$ have been exhibited by

G. W. Whitehead [156] and Zvengrowski [167]. Such a cross-section (following the latter) is given by $g : V_{8,3} \to S^7$, where

$$g(x, y, z) = -x(y^{-1}z).$$

Here the vectors $x, y, z \in R^8$ of the orthonormal 3-frame are treated as Cayley numbers once again. When $x = 1$, and so y and z are pure, this reduces to the cross-section $V_{7,2} \to V_{8,3}$ defined above. In the other direction we have

Proposition (1.16). _If_ $k \geq 1$ _then_ $V_{2k+1,k+1}$ _does not admit a cross-section over_ $V_{2k+1,k}$.

For suppose, on the contrary, that such a cross-section $f : V_{2k+1,k} \to V_{2k+1,k+1}$ exists. By (1.3) the restriction of f to $P_{2k+1,k}$ can be deformed into $P_{2k+1,k+1}$, yielding a map $h : P_{2k+1,k} \to P_{2k+1,k+1}$. Consider the induced homomorphism

$$h^* : H^r(P_{2k+1,k+1}) \to H^r(P_{2k+1,k}),$$

in mod 2 cohomology. Since f is a cross-section it follows that h^* is an isomorphism in dimension $2k$. However, the generator a_{2k} of $H^{2k}(P_{2k+1,k+1})$ is the cup-product square of the generator a_k of $H^k(P_{2k+1,k+1})$, which goes to zero since $H^k(P_{2k+1,k}) = 0$. Thus we obtain a contradiction which establishes (1.16).

Finally, let us fibre $V_{n,k}$ over $V_{n,l}$, where $1 \leq l < k < n$, by taking the last l vectors of each k-frame. Then (1.15) and (1.16) imply

Theorem (1.17). _Let_ $1 < l < k \leq n$. _Then_ $V_{n,k}$ _admits a cross-section over_ $V_{n,l}$ _if and only if_ (i) $l = k - 1$ _and_ (ii) _either_ $n = k$ _or_ $(n, k) = (7, 3)$ _or_ $(8, 4)$.

The cross-section of $V_{n,n}$ over $V_{n,n-1}$ is defined by taking g to be the exterior product. The other cases where a cross-section exists have been dealt with above. To establish the converse, note that a cross-section of $V_{n,k}$ over $V_{n,l}$ determines, by projection, a cross-section of $V_{n,l+1}$ over $V_{n,l}$ and determines, by restriction, a cross-section of $V_{n-1,k-1}$ over $V_{n-1,l-1}$. Using these observations and the results obtained above we arrive at (1.17).

2 · The Stiefel manifolds

We continue to denote the real numbers by R, the complex numbers by C, and the quaternions by H. Let A be one of these real division algebras, with the usual norm and operation of conjugation. An element $x \in A$ is called pure if $x + \bar{x} = 0$; in the real case only zero is pure. Complex numbers are regarded as pairs of real numbers, and quaternions as pairs of complex numbers, in the usual way.

Consider the free right module of dimension n over A $(n = 0, 1, \ldots)$ consisting of column vectors with entries in A. We denote this module by A^n or by nA, according to the context. The standard inner product $\langle\ ,\ \rangle$ is defined so that if

$$x = (x_1, \ldots, x_n), \quad y = (y_1, \ldots, y_n)$$

are elements of the module then

$$\langle x, y\rangle = \bar{x}_1 y_1 + \ldots + \bar{x}_n y_n.$$

The norm is given by $\|x\|^2 = \langle x, x\rangle$.

We embed A^n in A^{n+1} by adjoining zero as the last entry. The element with 1 in the rth place and zero everywhere else is denoted by a_r, and belongs to A^n for $n \geq r$.

Every complex n-vector $(z_1, \ldots, z_n)$ determines a real 2n-vector

$$(x_1, y_1, \ldots, x_n, y_n),$$

where $z_r = x_r + iy_r$, and similarly every quaternionic n-vector determines a complex 2n-vector. This convention is consistent with the embeddings described in the previous paragraph.

In A^n the vectors of norm ≤ 1 form a ball $B_n = B(A^n)$ of topological dimension dn, where $d = \dim_R A$, while the vectors of norm $= 1$ form a $(dn - 1)$-sphere $S_n = S(A^n)$. The norm-preserving automorphisms of A form a topological group $G_n = G(A^n)$, and G_n acts transitively on S_n so that the stabilizer of a_n is G_{n-1}. As usual we write $G(R^n) = O_n$, $G(C^n) = U_n$, $G(H^n) = Sp_n$. In the real case the subgroup of rotations is denoted by R_n rather than SO_n, to avoid confusion with the suspension functor. The embeddings

$$U_n \subset R_{2n}, \quad Sp_n \subset U_{2n}$$

are defined, in accordance with the conventions of the previous paragraph.

An ordered set $(u_1, \ldots, u_k)$ of vectors in A^n is called a k-frame. The k-frame is said to be non-singular if the vectors are linearly independent; this requires, of course, that $k \leq n$. The k-frame is said to be orthonormal if

$$\langle u_i, u_j \rangle = \delta_{ij} \quad (i, j = 1, \ldots, k),$$

where δ_{ij} is the Kronecker symbol. Note that a k-frame $(u_1, \ldots, u_k)$ in complex n-space determines a 2k-frame

$$(v_1, w_1, \ldots, v_k, w_k)$$

in real 2n-space, where v_r denotes the real vector determined by u_r, as above, and w_r denotes the real vector determined by $-iu_r$. If $(u_1, \ldots, u_k)$ is non-singular or orthonormal then so is $(v_1, w_1, \ldots, v_k, w_k)$. Similarly a quaternionic k-frame determines a complex 2k-frame, and the same remark applies.

After these preliminaries we are ready to define the Stiefel manifolds. To avoid trivialities, let $1 \leq k \leq n$. We topologize

$$A^n \times \ldots \times A^n \quad \text{(k factors)}$$

in the obvious way, and hence topologize the space of k-frames in A^n. The (non-compact) Stiefel manifold $O^*_{n,k}$ is defined as the subspace of non-singular k-frames and the (compact) Stiefel manifold $O_{n,k}$ as the subspace of orthonormal k-frames. Taking $k = 1$, for example, we see that $O^*_{n,1} = A^n - \{0\}$, $O_{n,1} = S_n = S(A^n)$.

In fact $O_{n,k}$ is always a deformation retract of $O^*_{n,k}$. To see this, recall the Gram-Schmidt process which associates to each non-singular k-frame $u = (u_1, \ldots, u_k)$ an orthonormal k-frame $u' = (u'_1, \ldots, u'_k)$. The process is continuous and has the property that the k-frame $(1 - t)u + tu' = ((1 - t)u_1 + tu'_1, \ldots, (1 - t)u_k + tu'_k)$ is non-singular, for $0 \leq t \leq 1$. Thus a deformation retraction $h_t : O^*_{n,k} \to O_{n,k}$ is given by $h_t(u) = (1 - t)u + tu'$. In view of this result we shall not consider $O^*_{n,k}$ any further but concentrate on the (compact) Stiefel manifold $O_{n,k}$.

Perhaps the easiest way to establish the smooth manifold structure is to consider $O_{n,k}$ as a G_n-space. Here the action of an element $g \in G_n$ transforms the k-frame $u = (u_1, \ldots, u_k)$ into $gu = (gu_1, \ldots, gu_k)$. The action of the group is transitive and the stabilizer of $(a_{n-k+1}, \ldots, a_n)$ is the subgroup $G_{n-k} \subset G_n$. Thus $O_{n,k}$ is homeomorphic to the factor space G_n/G_{n-k} of left cosets and thereby acquires its differential structure. In particular $O_{n,n} = G_n$, the group-space.

If $1 \leq l < k$ we can (see §7 of [133]) regard G_n/G_{n-k} as a fibre space over G_n/G_{n-l} with fibre G_{n-l}/G_{n-k}. In other words $O_{n,k}$ fibres over $O_{n,l}$ with fibre $O_{n-l,k-l}$. The projection p is given by taking the last l vectors of each k-frame to form an l-frame. The inclusion u of the fibre is given by adjoining to each $(k - l)$-frame in $(n - l)$-space the l-frame $(a_{k-l+1}, \ldots, a_k)$. If $l = 1$, in particular, then $O_{n,k}$ fibres over $O_{n,1} = S_n$ with fibre $O_{n-1,k-1}$. From this by a straightforward induction we obtain

Proposition (2.1). <u>The manifold</u> $O_{n,k}$ <u>has dimension</u> $\frac{1}{2}dk(2n - k + 1) - k$ <u>and connectivity</u> $d(n - k + 1) - 2$.

Further information can be obtained from the homotopy exact sequence of the fibration

$$\ldots \to \pi_r(O_{n-1,k-1}) \xrightarrow{u_*} \pi_r(O_{n,k}) \xrightarrow{p_*} \pi_r(S_n) \xrightarrow{\Delta} \pi_{r-1}(O_{n-1,k-1}) \to \ldots .$$

If m is odd the homotopy group $\pi_r(S^m)$ is finite for all $r > m$, as shown by Serre [125]. If m is even the same is true provided $r \neq 2m-1$. The order of the Whitehead square $w_m = [\iota_m, \iota_m]$ is infinite when m is even but after factoring out the sub-group of $\pi_{2m-1}(S^m)$ which w_m generates a finite group is obtained. These considerations lead to

Proposition (2.2). <u>In all cases $\pi_r(O_{n,k})$ is either finite or the direct sum of a finite and a cyclic infinite group. In the complex and quaternionic cases $\pi_r(O_{n,k})$ is finite for all even values of</u> r.

Following tradition we denote $O_{n,k}$ by $V_{n,k}$ in the real case, by $W_{n,k}$ in the complex case, and by $X_{n,k}$ in the quaternionic. The embeddings

$$W_{n,k} \subset V_{2n,2k}, \quad X_{n,k} \subset W_{2n,2k}$$

have already been defined. In the real case, as we have seen, there exist values of n, for any given k, such that the fibration $V_{n,k} \to V_{n,1}$ admits a cross-section. Are there analogous results in the complex and quaternionic cases? Since $W_{n,1} = V_{2n,1}$, a cross-section of $W_{n,k} \to W_{n,1}$ determines, by inclusion, a cross-section of $V_{2n,2k} \to V_{2n,1}$, and similarly a cross-section of $X_{n,k} \to X_{n,1}$ determines a cross section of $W_{2n,2k} \to W_{2n,1}$. Our next step is to introduce a construction, called the intrinsic join, which will enable us to prove some more general results on the existence of cross-sections.

The <u>join</u> $X * Y$ of spaces X and Y is the space obtained from the union of X, Y and $X \times Y \times I$ by identifying

$$(x, y, 0) = x, \quad (x, y, 1) = y \quad (x \in X, y \in Y).$$

A canonical homeomorphism $T : X * Y \to Y * X$ is given by $T(x, y, t) = (y, x, 1 - t)$. The join construction is functorial and can be relativized: thus if (Y, Z) is a pair we write

$$X * (Y, Z) = (X * Y, X * Z).$$

The join of spheres is again a sphere.

Consider the homeomorphism

$$h : S_m * S_n \to S_{m+n}$$

which is given by

$$h(x, y, t) = (x \cos \tfrac{\pi}{2} t, \ y \sin \tfrac{\pi}{2} t).$$

This can be generalized to a map

$$h = h_k : O_{m,k} * O_{n,k} \to O_{m+n,k}$$

as follows. Let $u = (u_1, \ldots, u_k)$ be a k-frame in m-space and $v = (v_1, \ldots, v_k)$ a k-frame in n-space. Then $h_k(u, v, t) = w$, where $w = (w_1, \ldots, w_k)$ is the k-frame in (m + n)-space given by $w_1 = h(u_1, v_1, t), \ldots, w_k = h(u_k, v_k, t)$. Clearly w is orthonormal if u and v are orthonormal. Also h_k is injective, for all values of k, and coincides with h for $k = 1$. Henceforth we write $h_k = h$, and refer to h as the <u>intrinsic map.</u> Note that

(2.3) $\quad ph = h(p * p)$,

as shown in the following diagram, where p denotes the standard projection from k-frames to l-frames and $1 \le l < k$.

$$\begin{array}{ccc} O_{m,k} * O_{n,k} & \xrightarrow{h} & O_{m+n,k} \\ \downarrow p * p & & \downarrow p \\ O_{m,l} * O_{n,l} & \xrightarrow{h} & O_{m+n,l} \end{array}$$

In particular take $l = 1$ and suppose that

$$f : O_{m,1} \to O_{m,k}, \quad g : O_{n,1} \to O_{n,k}$$

are cross-sections. Then (2.3) shows that

$$h(f * g)h^{-1} : O_{m+n,1} \to O_{m+n,k}$$

is also a cross-section; we shall refer to this as the <u>intrinsic join</u> of f and g. In the real case, for example, if f and g are Clifford cross-sections determined by Clifford modules M and N then the intrinsic join is the Clifford cross-section determined by $M \oplus N$. Thus the intrinsic join of Clifford cross-sections is again Clifford, also the intrinsic join of skew cross-sections is again skew, and the intrinsic join of homotopy-equivariant cross-sections is again homotopy-equivariant.

Given elements $\theta \in \pi_i(O_{m,k})$ $(i \geq 1)$, $\phi \in \pi_j(O_{n,k})$ $(j \geq 1)$, we can form the element $\theta * \phi \in \pi_{i+j+1}(O_{m,k} * O_{n,k})$, in the usual way, by taking the join of representative maps. If we now apply to $\theta * \phi$ the homomorphism induced by the intrinsic map

$$h : O_{m,k} * O_{n,k} \to O_{m+n,k}$$

we obtain an element $h_*(\theta * \phi) \in \pi_{i+j+1}(O_{m+n,k})$. We normally omit the h_*, since there is no risk of confusion with the ordinary join, and refer to the pairing of $\pi_i(O_{m,k})$ with $\pi_j(O_{n,k})$ to $\pi_{i+j+1}(O_{m+n,k})$ thus defined as the <u>intrinsic join.</u> The pairing is both bilinear and associative, as can easily be seen.

Think of the elements of $O_{n,k}$ as $n \times k$ matrices. By changing the sign of a column we obtain a self-map of class λ, say, and by changing the sign of a row we obtain a self-map of class μ. Of course $\lambda = 1$ and $\mu = 1$ except in the real case, which we have already discussed in §1. Clearly

$$(2.4) \quad \begin{cases} \text{(a)} \quad \lambda_*(\theta * \phi) = (\lambda_*\theta) * (\lambda_*\phi), \\ \text{(b)} \quad (\mu_*\theta) * \phi = \mu_*(\theta * \phi) = \theta * (\mu_*\phi). \end{cases}$$

Notice that $hT = Uh$, as shown below, where T is the switching map and U permutes the first m and last n row vectors.

$$\begin{array}{ccc} O_{m,k} * O_{n,k} & \xrightarrow{h} & O_{m+n,k} \\ T \downarrow & & \downarrow U \\ O_{n,k} * O_{n,k} & \xrightarrow[h]{} & O_{n+m,k} \end{array}$$

Hence we obtain the commutation law

$$(2.5) \quad \mu_*^{mn}(\theta * \phi) = (-1)^{(i+1)(j+1)}(\phi * \theta).$$

I emphasize that λ_* and μ_* are trivial apart from the real case.

From (2.3) and naturality of the ordinary join we obtain that

$$(2.6) \quad p_*(\theta * \phi) = (p_*\theta) * (p_*\phi),$$

as shown below, where p denotes the standard projection from k-frames to l-frames.

$$\begin{array}{ccc} \pi_i(O_{m,k}) \times \pi_j(O_{n,k}) & \xrightarrow{*} & \pi_{i+j+1}(O_{m+n,k}) \\ \downarrow p_* \times p_* & & \downarrow p_* \\ \pi_i(O_{m,l}) \times \pi_j(O_{n,l}) & \xrightarrow[*]{} & \pi_{i+j+1}(O_{m+n,l}) \end{array}$$

Let $\theta' \in \pi_i(O_{m,k-l})$, $\theta'' \in \pi_i(O_{m,l})$ denote the images of $\theta \in \pi_i(O_{m,k})$ under the homomorphisms induced by the standard projection, and let $\theta_\#$, $\theta'_\#$, $\theta''_\#$ be defined by taking the intrinsic join with θ, θ', θ'', respectively, as shown in the following diagram, where $t = i + j + 1$.

$$\begin{array}{ccccccccc} \cdots \to & \pi_j(O_{n-l,k-l}) & \to & \pi_j(O_{n,k}) & \to & \pi_j(O_{n,l}) & \to & \pi_{j-1}(O_{n-l,k-l}) & \to \cdots \\ & \downarrow \theta'_\# & & \downarrow \theta_\# & & \downarrow \theta''_\# & & \downarrow \theta'_\# & \\ \cdots \to & \pi_t(O_{m+n-l,k-l}) & \to & \pi_t(O_{m+n,k}) & \to & \pi_t(O_{m+n,l}) & \to & \pi_{t-1}(O_{m+n-l,k-l}) & \to \cdots \end{array}$$

Now (2.6) implies that the central square is commutative. It is not difficult to show, as in §2 of [67], that the other two squares are also commutative, sign-changes being tolerated as usual. Hence the given element θ determines the homomorphism of exact sequences $(\theta_\#, \theta'_\#, \theta''_\#)$. When $i = dm - 1$ and θ is the class of a cross-section this leads to

Theorem (2.7). *Suppose that $\theta \in \pi_{dm-1}(O_{m,k})$ is the class of a cross-section. Then*

$$\theta_\# : \pi_j(O_{n,k}) \to \pi_{j+dm}(O_{m+n,k})$$

is injective for $j < u$, surjective for $j \leq u$, where $u = 2d(n - k + 1) - 3$.

Corollary (2. 8). Suppose that $\theta \in \pi_{dm-1}(O_{m,k})$ and $\psi \in \pi_{dm+dn-1}(O_{m+n,k})$ are the classes of cross-sections, and that $n \geq 2k$. Then $\psi = \theta * \phi$, where $\phi \in \pi_{dn-1}(O_{n,k})$ is the class of a cross-section.

When $k = 1$ the homomorphism $\theta_{\#}$ is just iterated suspension, and the result is an immediate consequence of the Freudenthal theorem. Let $k \geq 2$ and suppose that (2. 7) is true with $k - 1$ in place of k. Then we obtain (2. 7) as stated by applying the five lemma to the homomorphism of exact sequences determined by θ, as in the diagram with $l = k - 1$. Thus (2. 7) follows by induction. We refer to this result as the generalized Freudenthal theorem.

There is a sense in which the transgression acts as a derivation with respect to the intrinsic join. Suppose that $1 \leq l < k \leq 2l$, with $m, n \geq k$ as before. Since $k - l \leq l$ the projection p from l-frames to $(k - l)$-frames is defined. Let $\theta \in \pi_i(O_{m,l})$, $\phi \in \pi_j(O_{n,l})$, where $i, j \geq 2$. Then

$$(2.9) \quad \Delta(\theta * \phi) = (\Delta\theta) * (p_*\phi) + (-1)^{i+1}(p_*\theta) * (\Delta\phi),$$

as indicated in the following diagram.

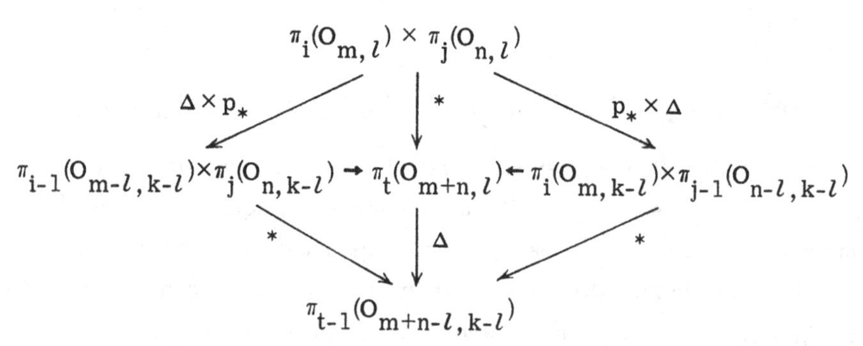

The proof of (2. 9) given in [67] is rather laborious and will be omitted, since we do not need to use it in what follows. Some other formulae involving the intrinsic join are given in §18 below.

3·The auxiliary spaces

The topological group $S = S_1$ acts on S_n and the orbit space $P_n = P(A^n)$ is the projective $(n - 1)$-space over A. Naturally P_{n+1} contains P_n, following our earlier embeddings, and the result of collapsing this subspace to a point is a sphere P_{n+1}/P_n of topological dimension dn. Since P_1 is a point-space the projective line P_2 is a d-sphere.

The identification of $S(C^n)$ with $S(R^{2n})$ described earlier induces a map $P(R^{2n}) = S(R^{2n})/S(R) \to S(C^n)/S(C) = P(C^n)$, which is a fibration with fibre the 1-sphere $S(C)/S(R)$. Similarly the identification of $S(H^n)$ with $S(C^{2n})$ determines a fibration

$$P(C^{2n}) = S(C^{2n})/S(C) \to S(H^n)/S(H) = P(H^n),$$

with fibre the 2-sphere $S(H)/S(C)$. We shall refer to these as the <u>standard fibrations.</u>

Let $P_{n,k}$, where $1 \leq k \leq n$, denote the space P_n/P_{n-k} obtained from P_n by collapsing P_{n-k} to a point. When $n = k$ the subspace is empty and we make the usual convention that $P_{n,n} = P_n^+$, the union of P_n and a point-space. We refer to $P_{n,k}$ as a <u>stunted projective space.</u> If, as is often convenient, we regard $P_{n,k}$ as an identification space of S_n directly then the point of $P_{n,k}$ corresponding to $(x_1, \ldots, x_n) \in S_n$ will be denoted by $[x_1, \ldots, x_n]$. Notice that $P_{n,1}$ is a sphere of topological dimension $d(n - 1)$. It follows that $P_{n,k}$, as a complex, has one cell e_r of dimension $d(r - 1)$ for $r = n - k + 1, \ldots, n$. Furthermore consider the cohomology ring $H^*(P_{n,k})$ with mod 2 coefficients in the real case, with integral coefficients in the complex and quaternionic cases. Recall that $H^*(P_n)$ is the truncated polynomial ring generated by a d-dimensional element σ with relation $\sigma^n = 0$. It follows that $H^*(P_{n,k})$, as a group, is generated by a set of elements $(\sigma_{n-k}, \ldots, \sigma_{n-1})$, where σ_r, for $n - k \leq r < n$, is the class carried by e_{r+1}. We make the convention that $\sigma_r = 0$ when $r \geq n$. Then $\sigma_r . \sigma_s = \sigma_{r+s}$ whenever $r, s \geq n - k$. The same notation etc. is used in mod p cohomology, for any prime p. Let $n - k \leq r < n$. If $p = 2$ then

$$(3.1) \quad Sq^{ds}\sigma_r = \binom{r}{s}\sigma_{r+s},$$

in all cases. If p is odd then

$$(3.2) \quad \mathcal{P}^s\sigma_r = \binom{r}{s}\sigma_{r+s(p-1)},$$

in the complex case, while

$$(3.3) \quad \mathcal{P}^s\sigma_r = \binom{2r}{s}\sigma_{r+s(p-1)/2}$$

in the quaternionic. These relations follow at once from the corresponding relations in the cohomology of the projective spaces which are readily established by the Cartan formula.

Although the relationship between Stiefel manifolds and stunted projective spaces is important, as we shall see, there is another family, the stunted quasiprojective spaces, which also plays a major role, particularly in the quaternionic case. This second auxiliary family is defined as follows.

The quasiprojective space $Q_n = Q_n(A)$ is defined to be the image of the map

$$\phi : S_n \times S \rightarrow G_n,$$

where $\phi(u, q)$ is the automorphism which leaves v fixed if $\langle u, v\rangle = 0$ and sends u to uq. Thus

$$\phi(u, q)v = u(q - 1)\langle u, v\rangle + v.$$

We can also define Q_n to be the space obtained from $S_n \times S$ by imposing the equivalence relation

$$(u, q) \sim (uz, z^{-1}qz) \qquad (z \in S),$$

and collapsing S_n to a point. It is easy to check that this construction yields a compact Hausdorff space and so can be identified with its embedded image in G_n. It is easy to check that $Q_n = P_n^+$ in the real case, and that Q_n is the suspension of P_n^+ in the complex case. However, Q_n is not the 3-fold suspension of P_n^+ in the quaternionic case, as we shall see.

We embed $Q_{n-k} \subset Q_n$ in the obvious way, so that $Q_{n-k} = Q_n \cap G_{n-k}$. We define the <u>stunted quasiprojective space</u> $Q_{n,k}$ to be the space obtained from Q_n by collapsing Q_{n-k} to a point. We also regard $Q_{n,k}$ as a subspace of $O_{n,k}$, in the obvious way. Note that $Q_{n,1} = O_{n,1}$.

When $1 \leq l < k$ we have cofibrations

$$P_{n-l,k-l} \to P_{n,k} \to P_{n,l},$$
$$Q_{n-l,k-l} \to Q_{n,k} \to Q_{n,l},$$

corresponding to the fibration

$$O_{n-l,k-l} \to O_{n,k} \to O_{n,l}.$$

The close relation between $O_{n,k}$ and $Q_{n,k}$ is already suggested by

Theorem (3.4). <u>The pair</u> $(O_{n,k}, Q_{n,k})$ <u>is</u> t-<u>connected, where</u> $t = 2d(n-k) + 3(d-1)$.

If $k = 1$ the assertion is trivial. Let $k \geq 2$ and suppose that the assertion is true with $k - 1$ in place of k. The inductive step and hence (3.4) will follow from

Theorem (3.5). <u>Let</u> $p : E \to B$ <u>be a fibration with fibre</u> F. <u>Let</u> $X \subset E$ <u>be a subspace such that</u> $p|X$ <u>is a cofibration with cofibre</u> $Y \subset F$. <u>Then</u>

$$\operatorname{conn}(E, X) \geq \min(\operatorname{conn}(F, Y), \operatorname{conn} B + \operatorname{conn} Y).$$

To establish this consider the diagram shown below, where i is the inclusion and $q = p|X$.

$$\begin{array}{ccc} \pi_r(X, Y) & \xrightarrow{q_*} & \pi_r(B) \\ {\scriptstyle i_*}\downarrow & & \downarrow{\scriptstyle 1} \\ \pi_r(E, F) & \xrightarrow[p_*]{} & \pi_r(B) \end{array}$$

It follows from the Blakers-Massey theorem, as stated in (1.25) of [143], that q_* is s-connected, where $s = \text{conn } B + \text{conn } Y$. Hence i_* is s-connected, since p_* is an isomorphism. Now (3.5) follows by the five lemma applied to the homomorphism of homotopy exact sequences induced by i, as shown below.

$$\begin{array}{ccccccccc} \cdots \to & \pi_r(Y) & \to & \pi_r(X) & \to & \pi_r(X, Y) & \to \cdots \\ & \downarrow & & \downarrow & & \downarrow & \\ \cdots \to & \pi_r(F) & \to & \pi_r(E) & \to & \pi_r(E, F) & \to \cdots \end{array}$$

Of course (3.4) gives some information about the cohomology of $O_{n,k}$ which is useful. In the complex case, for example, the information about cohomology operations obtained from (3.1), (3.2) implies

Proposition (3.6). <u>Suppose that</u> $W_{n,k}$ <u>admits a cross-section over</u> $W_{n,1}$. <u>Let</u> p <u>be any prime. Then</u> n <u>is divisible by the least power of</u> p <u>which exceeds</u> $(k - 1)/(p - 1)$.

Corollary (3.7). <u>Either suppose that</u> $W_{m,k}$ <u>admits a cross-section over</u> $W_{m,1}$, <u>where</u> $k \geq 3$, <u>or that</u> $X_{m,k}$ <u>admits a cross-section over</u> $X_{m,1}$, <u>where</u> $k \geq 2$. <u>Then</u> m <u>is divisible by</u> 3k <u>if</u> k <u>is a power of two, by</u> 4k <u>otherwise.</u>

Under the first hypothesis this follows immediately from (3.6). Under the second we use the cross-section of $X_{m,k}$ over $X_{m,1}$ to define, by inclusion, a cross-section of $W_{2m,2k}$ over $W_{2m,1}$, and then use the first part.

Consider the subspace $E_n \subset S_n$ consisting of all vectors x such that x_n is real and $x_n \geq 0$. Since $x_1, \ldots, x_{n-1}$ determines x_n, under these conditions, we see that E_n is a ball of topological dimension $d(n - 1)$. The identification map by which Q_n is defined determines, by restriction, a relative homeomorphism

$$h : (E_n \times S, E_n \times S \cup E_n \times 1) \to (Q_n, Q_{n-1}).$$

Hence Q_n can be obtained from Q_{n-1} by attaching a cell of dimension $dn - 1$. It follows by induction that Q_n is a complex having one cell

e_r of dimension $dr - 1$ for $r = 1, \ldots, n$. The first $n - k$ of these cells are contained in Q_{n-k}; hence the stunted quasiprojective space $Q_{n,k}$ contains one cell e_r for $r = n - k + 1, \ldots, n$. In cohomology with coefficients as before we find that $H^*(Q_{n,k})$, as a group, is generated by the classes τ_r $(r = n - k + 1, \ldots, n)$ carried by e_r. In the real case $Q_{n,k} = P_{n,k}$ and the multiplicative structure is as given before; in the other cases all products vanish. Now the action of G_n on $O_{n,k}$ determines, by restriction, a map

$$\phi : Q_n \times Q_{n,k} \to O_{n,k}.$$

It is not difficult to see, as in Chapter IV of [134], that if $r > s$ then ϕ maps $e_r \times e_s$ homeomorphically onto a cell $e_{r,s}$, say, of dimension $d(r + s) - 2$. Iterating this procedure we find eventually that for every sequence $J = (j_1, j_2, \ldots, j_t)$ of integers such that $n \geq j_1 > j_2 > \ldots > j_t > n - k$ the Stiefel manifold $O_{n,k}$ contains a cell e_J of dimension $d(j_1 + \ldots + j_t) - t$. The cells corresponding to different sequences are distinct. If we agree to associate the basepoint of $O_{n,k}$ with the empty sequence we obtain, in this way, a cellular decomposition of $O_{n,k}$, having $Q_{n,k}$ as a subcomplex. Of course this gives another proof of (3.4).

Cell-structures of this type were first described by Ehresmann [39], in relation to flag varieties. The construction was used by Miller [112] specifically for rotation groups, by J. H. C. Whitehead [158] for the real Stiefel manifold, and by Yokota [163] for the complex Stiefel manifold, who also published an incorrect account of the quaternionic case. The Steenrod version [134] covers all three cases and uses the information obtained from the cell-structure to determine the cohomology, with mod 2 coefficients in the real case and with integral coefficients in the complex and quaternionic cases. In what follows all we shall need is

Proposition (3.8). The cohomology ring $H^*(O_{n,k})$ (mod 2 coefficients in the real case, integral coefficients otherwise) is generated multiplicatively by a set of elements corresponding to the cells of $Q_{n,k}$.

The result is obvious for $k = 1$. Let $k \geq 2$, therefore, and assume the result is true with $k - 1$ in place of k. The Euler class of the fibration $O_{n,k} \to O_{n,k-1}$ vanishes, by (2.1), hence the Gysin sequence splits and so (3.8) follows by induction.

4 · Retractible fibrations

Let $p : E \to X$ be a fibration with fibre $F = p^{-1}(x_0)$, where $x_0 \in X$ is the base-point. We say that the fibration is <u>retractible</u> if there exists a map $\rho : E \to F$ such that $\rho | F \simeq 1$. If (E, F) is a CW-pair then ρ can, of course, be deformed into a retraction. Clearly the fibration is retractible if it is trivial, in the sense of fibre homotopy type. The Dold theorem shows (inter alia) that the converse holds under certain conditions. Following Dold [34] we begin by proving

Proposition (4.1). <u>Let</u> $p : E \to X$ <u>and</u> $p' : E' \to X$ <u>be fibrations and let</u> $f : E \to E'$ <u>be a fibre-preserving map. If</u> f, <u>as a map, is a homotopy equivalence then</u> f <u>is a fibre-homotopy equivalence.</u>

In this kind of proof it is convenient to allow homotopies over any interval $I_n = [0, n]$ $(n = 1, 2, \ldots)$. Let $k : E' \times I \to E'$ be a homotopy of ff' into 1, where $f' : E' \to E$ is a homotopy inverse of f. Lift $p'k$ to a homotopy $l : E' \times I \to E$ such that $l(e', 0) = f'e'$, and define $f'' : E' \to E$ by $f''e' = l(e', 1)$. Then f'' is fibre-preserving since $pf''e' = p'k(e', 1) = p'e'$. Now $ff'' \simeq 1$ by the (non-vertical) homotopy $h : E' \times I_2 \to E'$, where

$$\begin{aligned} h(e', t) &= fl(e', 1 - t) \qquad (0 \le t \le 1) \\ &= k(e', t - 1) \qquad (1 \le t \le 2). \end{aligned}$$

Note that $p'h$ is symmetric in the sense that $p'h(e', t) = p'h(e', 2 - t)$. Consider therefore the map $H : E' \times I_2 \times I \to X$ defined by

$$H(e', s, t) = \begin{cases} p'h(e', s) & (0 \le s \le 1 - t) \\ p'h(e', 1 - t) & (1 - t \le s \le 1 + t) \\ p'h(e', s) & (1 + t \le s \le 1). \end{cases}$$

Lift H to a map $K : E' \times I_2 \times I \to E'$ such that $K(e', s, 0) = h(e', s)$. Then $L : E' \times I_4 \to E'$ is a vertical homotopy of ff'' into 1, where

$$L(e', t) = \begin{cases} K(e', 0, t) & (0 \leq t \leq 1) \\ K(e', t - 1), 1) & (1 \leq t \leq 3) \\ K(e', 2, 4 - t) & (3 \leq t \leq 4). \end{cases}$$

Thus f admits a fibre homotopy right inverse which in turn, by the same argument, admits a fibre homotopy right inverse and that, for formal reasons, is fibre homotopic to f. Therefore f is a fibre homotopy equivalence, as asserted.

In the applications we have to deal with further conditions are satisfied which enable us to formulate the Dold theorem as

Theorem (4.2). *Suppose that* X *is path-connected, also that* E *and* E' *have the homotopy type of* CW-*complexes. Then a fibre-preserving map* $f : E \to E'$ *is a fibre homotopy equivalence if it induces a homotopy equivalence on the fibres.*

In fact it is sufficient to suppose that f induces a weak homotopy equivalence of the fibres. For by application of the five lemma to the induced homomorphism of the fibre homotopy sequence (with special treatment in the bottom dimensions) it follows that f itself is a weak homotopy equivalence and hence a homotopy equivalence, by the theorem of J. H. C. Whitehead [159]. Thus (4.2) follows from (4.1).

In particular, suppose that E is retractible with homotopy-retraction $\rho : E \to F$. Take E' to be the product fibration $X \times F$ and take f to be the fibre-preserving map which has ρ as its second component. Then (4.2) yields

Corollary (4.3). *Suppose that* X *is path-connected, also that* E *and* $X \times F$ *have the homotopy type of* CW-*complexes. If* E *is retractible then* E *is trivial, in the sense of fibre homotopy type.*

Let us say that the fibration is *decomposable* if E has the same homotopy type as $X \times F$. Does this condition imply that the fibration is trivial, in the sense of fibre homotopy type? This question is discussed in a recent article [81]. In general the answer is negative, but some positive results can be obtained as follows. We shall assume, for simplicity, that (E, F) is a CW-pair. We begin by proving

Theorem (4.4). Suppose that $F = S^q$, where $q \geq 1$, and $\pi_q(X) = 0$. If E is decomposable then the fibration is retractible.

For let $f : E \to X$, $g : E \to S^q$ be the components of a homotopy equivalence $E \to X \times S^q$. Since $\pi_q(X) = 0$ we have $g_* : \pi_q(E) \approx \pi_q(S^q)$ and $u_* \pi_q(S^q) = \pi_q(E)$, by exactness in the fibre homotopy sequence, where $u : F \subset E$. Hence $g_* u_* \pi_q(S^q) = \pi_q(S^q)$ and so the self-map gu of S^q has degree ± 1. Now (4.4) follows at once, and we deduce

Corollary (4.5). Let $n \geq k \geq 2$. If the fibration $p : O_{n,k} \to O_{n,k-1}$ is decomposable then it is trivial, in the sense of fibre homotopy type.

For (4.4) is applicable since
$q = d(n - k + 1) - 1 \leq d(n - k + 2) - 2 = \text{conn}(O_{n,k-1})$. We shall return to this in §20 below.

We say that a group G has the Hopf property if every homomorphism of G onto itself is an isomorphism. Finitely-generated abelian groups, for example, have the Hopf property. Following [88] we prove

Theorem (4.6). Suppose that $X = S^m$, where $m \geq 2$, and that $\pi_{m-1}(F)$ has the Hopf property. If E is decomposable then the fibration admits a cross-section.

For consider the fibre homotopy sequence

$$\ldots \to \pi_m(E) \xrightarrow{p_*} \pi_m(S^m) \xrightarrow{\Delta} \pi_{m-1}(F) \xrightarrow{u_*} \pi_{m-1}(E) \to 0.$$

If E is decomposable then $\pi_{m-1}(F) \approx \pi_{m-1}(E)$, since $\pi_{m-1}(S^m) = 0$. Since

$$u_* \pi_{m-1}(F) = \pi_{m-1}(E) \approx \pi_{m-1}(F)$$

it follows from the Hopf property that u_* is an isomorphism. Hence $\Delta \pi_m(S^m) = 0$ and $p_* \pi_m(E) = \pi_m(S^m)$, by exactness. Therefore E admits a cross-section, as asserted. We go on to prove

Theorem (4.7). Suppose that $X = S^m$, where $m \geq 2$, that $\pi_r(F)$ $(r = 1, 2, \ldots)$ has the Hopf property, and that $\pi_m(F)$ is finite.

If E is decomposable then the fibration is retractible.

For let $f : E \to S^m$, $g : E \to F$ be the components of a homotopy equivalence $h : E \to S^m \times F$. Let $f' : S^m \to E$, $g' : F \to E$ be defined by the restriction of a homotopy inverse $h' : S^m \times F \to E$. Then $ff' \simeq 1$, $gg' \simeq 1$ while gf', fg' are nulhomotopic. Now $p_* g'_* \pi_m(F) = 0$, since $\pi_m(F)$ is finite, and so

$$p_* f'_* \pi_m(S^m) = p_* \pi_m(E) = \pi_m(S^m),$$

by (4. 6). Therefore pf' is a map of degree ± 1, and so induces an automorphism of $\pi_r(S^m)$ for all r. Now if $\alpha \in \pi_r(F)$ then $p_* g'_* \alpha = p_* f'_* \beta$, for some $\beta \in \pi_r(S^m)$. Hence $g'_* \alpha - f'_* \beta \in u_* \pi_r(F)$, by exactness, while $g_*(g'_* \alpha - f'_* \beta) = \alpha$, since $gg' \simeq 1$ and gf' is nulhomotopic. Thus $g_* u_* \pi_r(F) = \pi_r(F)$ and so $g_* u_*$ is an automorphism of $\pi_r(F)$, by the Hopf property, for all values of r. Therefore gu is a homotopy equivalence, by the theorem of J. H. C. Whitehead [159]. By composing g with a homotopy inverse of gu we obtain a map $E \to F$ which is homotopic to a retraction. This proves (4. 7), and we deduce

Corollary (4. 8). In the complex and quaternionic cases, if the fibration $O_{n,k} \to S_n$ is decomposable then it is retractible and hence trivial in the sense of fibre homotopy type.

By (4. 6) the fibration has a cross-section and so $2n + 4 \geq 3k$, by (3. 7). Recall that $\pi_r(S^m)$ is finite when m is odd and $r > m$. Hence it follows by induction on k, using the homotopy exact sequence, that the homotopy groups $\pi(S_n, O_{n-1,k-1})$ are finite. Hence (4. 8) follows immediately from (4. 7).

This argument breaks down in the real case since the homotopy groups in question may be infinite. To overcome this difficulty we prove a cohomological counterpart of (4. 7), as follows. Consider the Wang exact sequence

$$\dots \to H^{r-m}(F) \to H^r(E) \xrightarrow{u_*} H^r(F) \to \dots ,$$

in mod 2 cohomology. If E is decomposable then

$$H^r(E) \approx H^r(F) \oplus H^{r-m}(F).$$

If $H^r(F)$ is finite for all r then u^* is surjective. Hence

Theorem (4.9). *Let* $p : E \to S^m$ *be a fibration with fibre* F, *where* $m \geq 2$. *Suppose that the* mod 2 *cohomology ring* $H^*(F)$ *is generated (multiplicatively) by elements of dimension less than* m. *If* E *is decomposable then* gu *induces an automorphism of* $H^*(F)$, *where* $g : E \to F$ *is as before.*

This can be exploited as follows. The fibre space E over S^m corresponds to an element $\alpha \in \pi_{m-1}(H, e)$ where H denotes the function-space of (free) self-maps of F, and e denotes the identity. Recall that E is trivial, in the sense of fibre homotopy type, if and only if $\alpha = 0$. Consider the homomorphism

$$c_\# : \pi_{m-1}(H, e) \to \pi_{m-1}(H, c),$$

induced by precomposition with a given self-map c. We prove

Lemma (4.10). *The map* $c : F \to F$ *can be extended over* E *if and only if* $c_\# \alpha = 0$.

Let $w_t : S^{m-1} \to B^m$ deform the inclusion into the constant map. The collapsing map $B^m \to S^m$ induces a trivial fibration over B^m. Hence there exists a relative homeomorphism

$$h : (B^m \times F, S^{m-1} \times F) \to (E, F),$$

such that the adjoint of $h|S^{m-1} \times F$ represents α. If c extends to a map $f : E \to F$ then

$$fh(w_t \times 1) : S^{m-1} \times F \to F$$

is a deformation rel $e \times F$ of $ch|S^{m-1} \times F \to F$ into $c\rho$, where $\rho : S^{m-1} \times F \to F$ is the projection. Hence $c_\# \alpha = 0$, by using the adjoint of the deformation. The argument is reversible and we obtain (4.10) as asserted.

These results enable us to prove

Theorem (4.11). <u>If the fibration</u> $V_{n+1,k+1} \to S^n$ <u>is decomposable then it is trivial, in the sense of fibre homotopy type.</u>

By (3.8) and (4.9) there exists a map $g : V_{n+1,k+1} \to V_{n,k}$ such that gu induces an automorphism of the mod 2 cohomology ring $H^*(V_{n,k})$. Hence it follows from the universal coefficient theorem that gu induces a $\mathcal{C}$-automorphism of the integral homology groups $H_r(V_{n,k})$, in all dimensions, where $\mathcal{C}$ denotes the class of abelian groups of odd order. Hence gu induces a $\mathcal{C}$-automorphism of $\pi_r(V_{n,k})$, in all dimensions, by the Serre-Hurewicz theorem as in [126]. Hence if $H_{n,k}$ denotes the space of self-maps of $V_{n,k}$ then

$$(gu)_\# : \pi_r(H_{n,k}, e) \to \pi_r(H_{n,k}, gu)$$

is a $\mathcal{C}$-isomorphism, by the comparison theorem, mod $\mathcal{C}$, applied to the spectral sequence of Federer [44]. The action of O_n on $V_{n,k}$ determines an embedding $i : O_n \subset H_{n,k}$. If $\sigma \in \pi_{n-1}(O_n, e)$ is the classifying element of our fibration in the sense of fibre bundle theory then $\tau = i_*\sigma \in \pi_{n-1}(H_{n,k}, e)$ is the classifying element in the fibre space sense. By (4.6) there exists a cross-section. Thus n is odd, hence $2\sigma = 0$, hence $2\tau = 0$. But $(gu)_\#\tau = 0$, by (4.10), and so the order of τ is odd. Therefore $\tau = 0$ and the proof of (4.11) is complete.

5 · Thom spaces

This section is partly based on the work of Atiyah [7]. Suppose, for simplicity, that X is a finite complex. The trivial n-plane bundle $R^n \times X$ is denoted by $n\underline{R}$ or simply by n. If U and V are (euclidean) vector bundles over X we say that U and V are J-<u>equivalent</u> (written $U \overset{J}{\simeq} V$) if there exist integers m and n such that the associated sphere-bundles $S(U \oplus m)$ and $S(V \oplus n)$ are of the same fibre homotopy type. The set of J-equivalence classes of vector bundles over X is denoted by $J(X)$, and the natural functor $\tilde{K}_R(X) \to J(X)$ by J. If U_1 and U_2 are vector bundles over X then $S(U_1 \oplus U_2)$ can be identified with the fibre join $S(U_1) * S(U_2)$. Now the fibre join of fibre homotopy equivalences is again a fibre homotopy equivalence. It follows that if $U_1 \overset{J}{\simeq} V_1$ and $U_2 \overset{J}{\simeq} V_2$ then $U_1 \oplus U_2 \overset{J}{\simeq} V_1 \oplus V_2$. Hence $J(X)$ acquires an abelian group structure from the direct sum, making J a homomorphism. We now prove

Theorem (5.1). <u>The group</u> $J(X)$ <u>is finite.</u>

Since $\tilde{K}_R(X)$ is finitely generated it is sufficient to show that $J(X)$ is a torsion group. In [7] Atiyah proves this using the classifying space for fibrations but the following argument seems simpler. It is sufficient to prove

Lemma (5.2). <u>Let</u> $E = S(V)$ <u>be a sphere-bundle over</u> X. <u>Then</u> $E_m = S(mV)$ <u>is retractible, for some positive integer</u> m.

Since V can be replaced by a Whitney multiple of itself, if necessary, we can suppose without real loss of generality that $q > \dim X$, where $q = \dim V - 1$, and that V is oriented. If $\dim X \leq 1$ then (5.2) is trivial. Let Y be a subcomplex of X such that $X = e^n \cup Y$, where $n \geq 2$, and make the inductive hypothesis that (5.2) is true with Y in place of X. Without real loss of generality we can suppose that $E|Y$ is retractible. Consider the relative homeomorphism $f : (B^n, S^{n-1}) \to (X, Y)$ by which the n-cell e^n is attached. The induced bundle f^*E is a product, and so f can be lifted to a fibre-preserving relative homeomorphism

$$g : (B^n \times S^q, S^{n-1} \times S^q) \to (E, F),$$

where $F = E|Y$ and S^q is the fibre. Thus $E - F$ is the union of an n-cell, the image of $(B^n - S^{n-1}) \times e$ under g, and an $(n+q)$-cell, the image of $(B^n - S^{n-1}) \times (S^q - e)$. Choose a retraction $\rho : F \to S^q$ and consider the composition $\rho h : S^{n-1} \times S^q \to S^q$, where $h = g|S^{n-1} \times S^q$. We can extend ρh over $B^n \times e \cup S^{n-1} \times S^q$, since $n < q$, and then the obstruction to further extension over $B^n \times S^q$ is an element θ, say, of $\pi_{n+q-1}(S^q)$. If we take the r-fold fibre join of this situation with itself, where $r \geq 2$, then θ is replaced by θ_r, say, where θ_r is stably equivalent to $r\theta$ (see Part I of [4]). Since the stable group of the $(n-1)$-stem is finite we can choose r so that $\theta_r = 0$, and hence the r-fold fibre-join of ρh with itself can be extended over the r-fold fibre join of f^*E with itself. By composing such an extension with the inverse of the r-fold fibre join of g with itself, we obtain a retraction of $S(rV)$. Therefore (5.2) follows by induction and so (5.1) is proved.

The argument we have just given is used by Adams in Part I of [4] to establish an important generalization of the lemma, known as the 'Dold theorem mod k'. In this the hypothesis is that there exists a map $\rho : E \to S^q$ such that $\rho|S^q$ has degree k, and the conclusion is that m, in (5.2), can be chosen to be a power of k.

Let V, as before, be a (euclidean) vector bundle over the finite complex X. The vectors $v \in V$ such that $|v| \leq 1$ form the associated ball bundle $B(V)$, while those such that $|v| = 1$ form the associated sphere bundle $S(V)$. The Thom space X^V is defined to be the space obtained from $B(V)$ by collapsing $S(V)$ to a point. If the fibre of V is identified with R^n, where $n = \dim V$, and B^n/S^{n-1} with S^n, then S^n is embedded in X^V. We say that X^V is retractible (alternatively coreducible) if S^n is a retract of X^V. This is the case, for example, if $S(V)$ is retractible in the sense of §4. Notice that

(5.3) $\text{conn}(X^V, S^n) = n + \text{conn } X$.

Following Bott [31] we prove

Theorem (5.4). If X^V is retractible then so is $S(V \oplus 1)$.

We can obtain $S(V \oplus 1)$ from $B(V)$ by identifying points of $S(V)$ with their images under the projection $S(V) \to X$. Thus X^V can be obtained from $S(V \oplus 1)$ by collapsing the canonical cross-section to a point. Note that S^n is mapped identically, where $n = \dim V$. By composing a retraction $X^V \to S^n$ with the collapsing map $S(V \oplus 1) \to X^V$ we obtain a retraction $S(V \oplus 1) \to S^n$, as required. In fact $S(V)$ itself is retractible, by suspension theory, if $n - 1 > \dim X$.

The homotopy type of X^V is determined by the homotopy type of the pair $(B(V), S(V))$ and hence, using the fibre-cone construction, by the fibre homotopy type of $S(V)$. If $U = V \oplus m$, where $m \geq 1$, then X^U is homeomorphic to $S^m X^V$. Thus the S-type of X^V depends only on the class α, say, of V in $J(X)$, and can therefore be written as X^α. We say that X^α is retractible if X^V is S-retractible. Thus (5.4) implies

Corollary (5.5). If X^α is retractible, where $\alpha \in J(X)$, then $\alpha = 0$.

Since J is a contravariant functor we can consider the orbits of $J(X)$ under $G(X)$, the group of homotopy classes of homotopy equivalences of X with itself. If two elements lie in the same orbit then their Thom spaces have the same S-type. Under certain conditions Feder and Gitler [43] have shown that this correspondence between orbits and S-types is bijective, and have used this in [42], [43] to classify stunted projective spaces by S-type. The classification is complete in the complex and quaternionic cases but not in the real case. However, I understand that Gitler and Mahowald, using earlier results of Mahowald [100], have recently succeeded in completing the classification in the real case as well.

If U, V are euclidean bundles over X then the Thom spaces of U and $U \oplus V$ are related as follows. Let $q : S(V) \to X$ denote the projection in the associated sphere-bundle. Consider the inclusion

$$\frac{B(U) \times S(V)}{S(U) \times S(V)} \to \frac{B(U) \times B(V)}{S(U) \times B(V)},$$

where $\times$ means the fibre product over X. Here the domain is the Thom space of the induced bundle q^*U over $S(V)$ while the codomain contains the Thom space of U itself as a deformation retract. Collapsing the subspace to a point yields

$$\frac{B(U) \times B(V)}{B(U) \times S(V) \cup S(U) \times B(V)} = \frac{B(U \oplus V)}{S(U \oplus V)},$$

the Thom space of $U \oplus V$. In this sense, therefore, we have a cofibration of the form

$$(5.6) \quad S(V)^{q^*U} \to X^U \to X^{U \oplus V}.$$

We now show, following Atiyah [7], that the stunted projective spaces and stunted quasi-projective spaces of §3 can be obtained as Thom spaces of bundles over projective spaces. Recall that the Hopf (or canonical) line bundle $L = L_A$ over $P(A^k)$ is obtained from $A \times S(A^k)$ by identifying (u, v) with (uz, vz) for all $z \in S$, where $u \in A$, $v \in S(A^k)$. Hence the ball bundle associated with $(n - k)L$, where $n \geq k$, is the space obtained from $B_{n-k} \times S_k$ by identifying (u, v) with (uz, vz) for all $z \in S$, where $u \in B_{n-k}$, $v \in S_k$. An equivariant relative homeomorphism

$$(5.7) \quad h : (B_{n-k} \times S_k, S_{n-k} \times S_k) \to (S_n, S_{n-k})$$

is given by $h(u, v) = (u, (1 - t^2)^{\frac{1}{2}}v)$, where $t = |u|$. Factoring out the action of S we obtain a homeomorphism between the Thom space of $(n - k)L$, over P_k, and the stunted projective space $P_{n,k}$.

The stunted quasi-projective spaces can be similarly represented, with the help of a certain (real) $(d - 1)$-plane bundle L' over P_k. To define this, let $A' \subset A$ be the subspace of pure elements u (i.e. those such that $\bar{u} = -u$) and let S act on A' so that $z \in S$ sends u into $z^{-1}uz$. Then L' is obtained from $A' \times S(A^k)$ by factoring out the diagonal action. The same construction as was used in the earlier case, with obvious modification, yields a homeomorphism between the Thom space of $(n - k)L \oplus L'$, over P_k, and the stunted quasi-projective space $Q_{n,k}$. We omit the details. In the real case $L' = \underline{0}$, and can be ignored. In the complex case $L' = \underline{1}$ and so $Q_{n,k}(C)$ is the suspension of $P_{n,k}(C)$, as can easily be seen directly. In the quaternionic case L' is the 3-plane bundle associated with the standard fibration $g : P(C^{2k}) \to P(H^k)$. Hence it follows from (5.6) that $Q_{n,k}(H)$ is the cofibre of the map

$$P_{2n,2k}(C) \to P_{n,k}(H)$$

induced by the standard fibration.

When the base space is a sphere the Thom space can be described in terms of the Hopf construction, as follows. Given a map $T : S^{m-1} \to O_n$, of homotopy class $\alpha \in \pi_{m-1}(O_n)$, we first define

$$f : S^{m-1} \times S^{n-1} \to S^{n-1},$$

where $f(x, y) = T(x)(y)$ $(x \in S^{m-1},\ y \in S^{n-1})$. The sphere-bundle $S(V)$ over S^m which corresponds to T can be obtained from $(B^m \times S^{n-1}) + S^{n-1}$ by identifying points of $S^{m-1} \times S^{n-1}$ with their images under f, as explained in [89]. Hence it follows that the Thom space of V can be obtained from $(B^m * S^{n-1}) + S^n$ by identifying points of $S^{m-1} * S^{n-1}$ with their images under

$$h : S^{m-1} * S^{n-1} \to S^n,$$

where h is obtained from f by the Hopf construction. In terms of homotopy type, therefore, the Thom space is the mapping cone of $J\alpha \in \pi_{m+n-1}(S^n)$, the homotopy class of h. Note that the Thom space is retractible if and only if $J\alpha = 0$.

For example, consider $P_{n,2}$ and $Q_{n,2}$, regarded as Thom spaces of vector bundles over the d-sphere P_2. In the complex case the Hopf line bundle is classified by the generator of $\pi_1(O_2)$, which corresponds under J to the Hopf class in $\pi_3(S^2)$. We see, therefore, that $Q_{n,2} = SP_{n,2}$ is the mapping cone of $(n - 2)\eta$, where $\eta = \eta_{2n-3}$ is the $(2n - 5)$-fold suspension of the Hopf class. It follows at once that

(5.8) $\Delta\iota_{2n-1} = (n-2)\eta$,

where Δ denotes the transgression operator in the homotopy sequence of the fibration $W_{n,2} \to W_{n,1}$. Hence a cross-section exists if and only if n is even.

In the quaternionic case we are concerned with euclidean m-plane bundles over S^4, which are classified by elements of $\pi_3(O_m)$. Points of S^3 are represented by quaternions of unit norm and points of S^2 by pure quaternions of unit norm. Let

$$O_3 \xleftarrow{a} S^3 \xrightarrow{b} O_4$$

be defined by

$$a(z)(q) = z^{-1}qz, \quad b(z)(r) = rz,$$

where $z, r \in S^3$ and $q \in S^2$. The 3-plane bundle L' over P_2 described above corresponds to a, under the standard classification, while the 4-plane bundle L corresponds to b. If $\alpha \in \pi_3(O_3)$ denotes the class of a then $J\alpha = \omega$, the Blakers-Massey generator of $\pi_6(S^3) = Z_{12}$. If $\beta \in \pi_3(O_4)$ denotes the class of b then $J\beta$ is the Hopf class in $\pi_7(S^4)$. Thus $P_{n,2}$ is the mapping cone of $(n-2)\nu$, where $\nu = \nu_{4n-8}$ denotes the (4n - 12)-fold suspension of the Hopf class. Also $Q_{n,2}$ is the mapping cone of ω, for $n = 3$, and the mapping cone of $n\nu$, for $n > 3$, since ω is stably equivalent to 2ν. It follows at once that

$$(5.9) \quad \Delta\iota_{4n-1} = \begin{cases} \omega & (n = 2) \\ n\nu & (n > 2), \end{cases}$$

where Δ denotes the transgression operator for the fibration $X_{n,2} \to X_{n-1}$. In the stable range ν is of order 24. Hence a cross-section exists if and only if n is a multiple of 24. Moreover $P_{n,2}(H)$ and $Q_{n,2}(H)$ are never of the same S-type.

This information about the quaternionic case can be used to prove

Proposition (5.10). _If_ $m \equiv 0 \bmod 24$ _then_ $W_{m,4}$ _admits a cross-section over_ $W_{m,1}$.

For all $m > 4$ the group $\pi_{2m-2}(W_{m-1,2})$ is an extension of the cokernel of $Z_2 = \pi_{2m-1}(W_{m-1,1}) \xrightarrow{\Delta} \pi_{2m-2}(W_{m-2,1}) = Z_{24}$ by the cokernel of

$$Z_2 = \pi_{2m-2}(W_{m-1,1}) \xrightarrow{\Delta} \pi_{2m-3}(W_{m-2,1}) = Z_2.$$

When m is even both these transgression homomorphisms are injective, by (5. 8), and so $\pi_{2m-2}(W_{m-1,2}) = Z_{12}$. Hence for $m \equiv 0 \bmod 24$ it follows from (5. 9) and naturality that

$$\Delta : \pi_{2m-1}(W_{m,1}) \to \pi_{2m-2}(W_{m-1,2})$$

is trivial and hence that $W_{m,3}$ admits a cross-section over $W_{m,1}$, by exactness. However these cross-sections of $W_{m,3}$ can be lifted to $W_{m,4}$, as is shown by the exact sequence

$$\pi_{2m-1}(W_{m,4}) \xrightarrow{p_*} \pi_{2m-1}(W_{m,3}) \to \pi_{2m-2}(W_{m-3,1}) = 0.$$

6 · Homotopy-equivariance

The purpose of this section is to introduce a series of homotopy-G notions, where G is a discrete group. For any G-space X we denote by $g_{\#} : X \to X$ the action of an element $g \in G$. If X and Y are G-spaces we describe a map $f : X \to Y$ as a homotopy-G map if

$$g_{\#} f g_{\#}^{-1} \simeq f,$$

for all $g \in G$. If f is a homotopy-G map and a homotopy equivalence then any homotopy inverse of f is also a homotopy-G map; in that case we describe f as a homotopy-G equivalence and say that X and Y have the same homotopy-G type. Also we describe a G-space X as (homotopy-G) neutral if $g_{\#} \simeq 1$ for all $g \in G$. This is the case, for example, if X is contractible with any G-structure, or if X is a sphere with orientation-preserving G-structure.

Now consider the category of G-spaces E, F, ... over a given G-space X. We describe a map $f : E \to F$ over X as a homotopy-G overmap if $g_{\#} f g_{\#}^{-1}$ is homotopic to f over X for all $g \in G$. When E, F, ... are fibre spaces over X the term fibre-preserving homotopy-G map may be used instead. The other homotopy-G notions are extended to the category of spaces over X in the obvious way.

Let U, V be G-vector bundles over the G-space X. We say that an isomorphism $f : U \to V$ of vector bundles is a homotopy-G isomorphism if f and $g_{\#} f g_{\#}^{-1}$ are homotopic through isomorphisms, for all $g \in G$. If such an isomorphism exists we say that U and V are homotopy-G isomorphic. Notice that if U, V and W are G-vector bundles over X with U homotopy-G isomorphic to V then $U \oplus W$ is homotopy-G isomorphic to $V \oplus W$ and $U \otimes W$ is homotopy-G isomorphic to $V \otimes W$.

From now on we suppose that X is a pointed G-space, i.e. $g_{\#}$ is a pointed map for all $g \in G$. The Thom space of a G-vector bundle

V over X is a pointed G-space X^V. If $U = V \oplus n\underline{R}$, where $n \geq 1$, then X^U is G-equivalent to $S^n X^V$. If $S(V)$ is retractible, as a homotopy-G space, then so is X^V; conversely if X^V is retractible then so is $S(V \oplus 1)$, by the argument used in §5. Moreover if U and V are G-vector bundles such that $S(U)$ and $S(V)$ have the same fibre homotopy-G type then X^U and X^V have the same homotopy-G type. Following [82] we say that U and V are J/G-equivalent if $S(U \oplus m\underline{R})$ and $S(V \oplus n\underline{R})$ have the same fibre homotopy-G type, for some m and n, where $\underline{R}$ has trivial G-structure.

The homotopy-G version of Spanier-Whitehead S-theory presents no difficulty. The suspension of a homotopy-G map is a homotopy-G map, and the converse holds in the stable range. Thus the notions of stable homotopy-G type, etc. are defined. Notice that the J/G-equivalence class of a G-vector bundle determines the stable homotopy-G type of the Thom space. It is not hard to show (see [82]) that the J/G-order is always finite.

For example, let $G = Z_2$ act on the real projective space P^{k-1} by reflection in the hyperplane of the last coordinate. Regard the Hopf line bundle L over P^{k-1} as a Z_2-vector bundle in the obvious way. Then the Thom space of $(n - k)L$ corresponds, under the homeomorphism of §5, to the stunted real projective space $P_{n,k}$, with Z_2 still acting by reflection in the hyperplane of the last coordinate. Thus the stable homotopy-Z_2 type of $P_{n,k}$, with this Z_2-structure, depends only on the residue class of $n \bmod \hat{a}_k$, where $\hat{a}_k$ denotes the J/Z_2-order of L over P^{k-1}.

The treatment of duality requires a brief comment. Let $f : X \wedge Y \to S^n$ be a duality map in the ordinary sense (see [131]). If X and Y are G-spaces we describe f as a homotopy-G duality map if f is a homotopy-G map with respect to the neutral G-structure on S^n. In that case the dual of the stable homotopy class of $g_\# : X \to X$ is the stable homotopy class of $g_\#^{-1} : Y \to Y$. Thus Y is stable neutral, or S-neutral, if and only if X is.

From now on take $G = Z_2$, for simplicity; the general case is dealt with in [82]. Then $g_\#$ is an involution, where G is the generator. Suppose that we have a Z_2-vector bundle V over the Z_2-space X which is trivial as a vector bundle. Choose a trivialization $\theta : V \to X \times M$, where M is a trivial Z_2-module, and transfer the Z_2-structure of V to $X \times M$ through θ. Use the automorphism of $X \times M$ thus obtained as a clutching function and thereby construct a vector bundle V_θ over the suspension SX. Clearly θ is a homotopy-Z_2 isomorphism if and only if V_θ is trivial as a vector bundle. Furthermore

$$S(\theta) : S(V) \to X \times S(M)$$

is a fibre homotopy-Z_2 equivalence if and only if $S(V_\theta)$ is trivial in the sense of fibre homotopy type. If we replace θ by ϕ, say, where ϕ is also a vector bundle trivialization of V, then $V_\theta \oplus U$ is stably isomorphic to $V_\phi \oplus (Sg_\#)^*U$, where U is the vector bundle constructed by treating the automorphism $\phi\theta^{-1}$ of $X \times M$ as a clutching function.

Recall that the $\tilde{K}_R$-order of L over P^{k-1} is precisely a_k, the power of 2 defined in §1. Regarding P^{k-1} as a Z_2-space and L as a Z_2-vector bundle we now prove

Proposition (6.1). _If $k > 2$ and $k \not\equiv 0 \bmod 4$ then $a_k L$ is trivial, in the sense of homotopy-Z_2 isomorphism. If $k = 2$ or $k \equiv 0 \bmod 4$ then $a_k L$ is non-trivial, in the same sense, while $2a_k L$ is trivial._

Take $k = 2$, to start with, and identify $P^1 = S^1$, in the usual way, so that Z_2 acts by reflection on S^1. Any trivialization of 2L, as a vector bundle, yields a generator $\gamma \in \tilde{K}_R(S^2) = Z_2$, through the construction we have just described. Hence 2L is stably non-trivial, as a Z_2-vector bundle. Moreover 2L is not J/Z_2-trivial, since $J\gamma \neq 0$. The proof that 4L is trivial in the sense of homotopy-Z_2 isomorphism is left as an exercise.

To deal with the general case, let us denote the real line by R or R' according as Z_2 acts trivially or not, so that

$$P^{k-1} = P((k-1)R \oplus R')$$

as a Z_2-space. In the Clifford algebra C_k let $e = e_k$ denote the last element of the generating set. The action of Z_2 on the sphere

$$S^{k-1} = S((k-1)R \oplus R')$$

is given by $x \mapsto e.x.e$, where . denotes multiplication in the Clifford algebra. We regard C_k as a Z_2-graded algebra in the usual way (see [9]). Let (M^0, M^1) be a graded module over C_k with $\dim M^0 = n$, say. The vector bundle nL, over P^{k-1} as a space, can be identified with $(S^{k-1} \times M^0)/Z_2$ where Z_2 acts by sign-reversal on both S^{k-1} and M^0. Under this identification, the involution on the Z_2-vector bundle nL, over P^{k-1} as a Z_2-space, transforms $\pm(x, y)$ into $\pm(e.x.e, y)$. Consider the vector bundle trivialization

$$\theta : (S^{k-1} \times M^0)/Z_2 \to P^{k-1} \times M^1$$

given by $\theta(\pm(x, y)) = (\pm x, x.y)$. This transforms the involution on the domain into the involution on the codomain which sends (x, z) into $(x, \psi(x, z))$, where $\psi : P^{k-1} \times M^1 \to M^1$ is given by $\psi(\pm x, z) = x.e.x.e.z$ $(x \in S^{k-1}, z \in M^1)$. Now ψ is equal to the composition

$$P^{k-1} \times M^1 \xrightarrow{\pi \times \sigma} S^{k-1} \times M^0 \xrightarrow{\mu} M^1,$$

where $\pi(\pm x) = x.e.x$, $\sigma z = e.z$, and $\mu(x, y) = x.y$. Hence $(nL)_\theta \approx (S\pi)^*W$, where $(nL)_\theta$ is the vector bundle over SP^{k-1} obtained from ψ by the clutching construction and W is the vector bundle over S^k obtained from μ by the clutching construction. If (M^0, M^1) is irreducible, so that $n = a_k$, then $[W]$ generates $\tilde{K}_R(S^k)$, as shown in [9]. Now $\tilde{K}_R(S^k) = 0$ when $k \equiv 3, 5, 6$ or $7 \bmod 8$; moreover $(S\pi)^*\tilde{K}_R(S^k) = 0$ when $k > 2$ and $k \equiv 1$ or $2 \bmod 8$ (see Karoubi [90] for example). This proves (6.1) when $k > 2$ and $k \not\equiv 0 \bmod 4$.

This is no longer true when $k \equiv 0 \bmod 4$. To see this consider the following exact sequence, where $u : P^{k-2} \subset P^{k-1}$.

$$\tilde{K}_R(S^k) \xrightarrow{(S\pi)^*} \tilde{K}_R(SP^{k-1}) \xrightarrow{(Su)^*} \tilde{K}_R(SP^{k-2})$$

When $k \equiv 0 \bmod 4$ we have $\tilde{K}_R(S^k) = Z$ while $\tilde{K}_R(SP^{k-1}) = Z \oplus Z_2$, $\tilde{K}_R(SP^{k-2}) = Z_2$, as shown by Karoubi [90]. By elementary algebra the image $(S\pi)^*\gamma \in \tilde{K}_R(SP^{k-1})$ of a generator $\gamma \in \tilde{K}_R(S^k)$ cannot be halved, and so we obtain that a_kL is stably non-trivial, in the sense of homotopy-

Z_2-isomorphism. Of course $2a_kL$ is trivial, as a Z_2-vector bundle, since $2a_k = a_{k+1}$. In the next section we shall show that a_kL is not J/Z_2-trivial when $k \equiv 0 \bmod 4$.

Returning to the general case we now describe another approach, due to Atiyah and Segal, which will be needed later. For any Z_2-space X let $\hat{X}$ denote the <u>mapping torus</u> of $g_\#$, i.e. the space obtained from $X \times I$ by identifying $(x, 0)$ with $(g_\# x, 1)$ for all $x \in X$. If X is neutral, for example, then $\hat{X}$ has the homotopy type of $X \times S^1$. For any Z_2-vector bundle V over X we regard $\hat{V}$ as a vector bundle over $\hat{X}$ in the obvious way. Let U, V be Z_2-vector bundles over X and let $f : U \to V$ be a homotopy-Z_2 isomorphism. Thus there exists a homotopy h_t of f into $g_\# f g_\#^{-1}$ which is an isomorphism for all values of t. Hence an isomorphism $\hat{f} : \hat{U} \to \hat{V}$ is defined by

$$\hat{f}(u, t) = (h_t u, t) \qquad (u \in U, t \in I).$$

Therefore $\hat{U}$ and $\hat{V}$ are (stably) isomorphic if U and V are (stably) homotopy-Z_2 isomorphic. Similarly $\hat{U}$ and $\hat{V}$ are J-equivalent if U and V are J/Z_2-equivalent. The converse is also true but will not be needed in what follows.

To apply this <u>principle of the mapping torus</u>, some information about $\tilde{K}_R(\hat{X})$ is required. This can be obtained by various ad hoc methods, as we shall see, or by regarding $\hat{X}$ as a fibre bundle over S^1 with fibre X and 'monodromy' exact sequence of the form

$$\ldots \to \tilde{K}_R(SX) \to \tilde{K}_R(\hat{X}) \to \tilde{K}_R(X) \xrightarrow{1 - g_\#^*} \tilde{K}_R(X).$$

7 · Cross-sections and the S-type

Following Woodward [161] we begin by proving

Theorem (7.1). If the Stiefel manifold $O_{n,k}$ admits a cross-section then the sphere-bundle $S(nL)$ is trivial as a fibre space over P_k.

Recall that the total space of $S(nL)$ is obtained from $S_n \times S_k$ by identifying

$$(u, v) \sim (uz, vz) \qquad (u \in S_n, \ v \in S_k),$$

for all $z \in S$. Given a map $f : S_n \to O_{n,k}$ we define a self-map f' of $S_n \times S_k$ by

$$f'(u, v) = (f_u(v), v),$$

where $f_u : S_k \to S_n$ denotes the S-map given by $f(u)$. Factoring out the action of S we see that f' induces a fibre-preserving map

$$f'' : S(nA) \to S(nL),$$

over P_k. If f is a cross-section then f'' is a homeomorphism over the point space P_1 and so f'' is a fibre homotopy equivalence, by Dold's theorem. This proves (7.1) and hence the 'only if' part of

Theorem (7.2). Let $t_k > 0$ be the J-order of the Hopf line bundle over P_k. Then $O_{n,k}$ admits a cross-section if and only if $n \equiv 0 \mod t_k$.

To prove the rest, consider the function-space $E_{n,k}$ of S-equivariant maps $S_k \to S_n$. Since every element of $O_{n,k}$ determines such a map we can regard $O_{n,k}$ as a subspace of $E_{n,k}$. Note that

$O_{n,1} = E_{n,1}$, in particular. We now prove

Theorem (7.3). <u>The pair</u> $(E_{n,k}, O_{n,k})$ <u>is</u> t-<u>connected, where</u> $t = 2d(n - k + 1) - 3$.

In the real case this result is due to Haefliger and Hirsch [52]; the general case is dealt with in [73] by essentially the same method. The first step is to show that the restriction map $E_{n,k} \to E_{n,k-1}$ is a fibration. Taking the inclusion map $S_{k-1} \subset S_n$ as basepoint in $E_{n,k-1}$ we see that the fibre $F_{n,k}$ of the fibration contains the fibre S_{n-k+1} of the fibration $O_{n,k} \to O_{n,k-1}$. The second step is to show that the pair $(F_{n,k}, S_{n-k+1})$ has the same homotopy type as the pair $(\Omega^{d(k-1)}S_n, S_{n-k+1})$, where Ω denotes the space of loops and the embedding is the standard one. The last step is to consider the homomorphism of exact sequences shown below and proceed by induction on k.

$$\begin{array}{ccccccc} \dots \to \pi_r(S_{n-k+1}) & \to & \pi_r(O_{n,k}) & \to & \pi_r(O_{n,k-1}) \to \dots \\ \downarrow & & \downarrow & & \downarrow \\ \dots \to \pi_r(F_{n,k}) & \to & \pi_r(E_{n,k}) & \to & \pi_r(E_{n,k-1}) \to \dots \end{array}$$

The vertical homomorphism on the left is equivalent to the iterated suspension

$$\pi_r(S_{n-k+1}) \to \pi_{r+d(k-1)}(S_n),$$

by the second step, and so is an isomorphism, by the Freudenthal theorem in the relevant range; hence the inductive step follows by a five lemma argument and we obtain (7.3).

We are now ready to complete the proof of (7.2). By hypothesis, $S(nL)$ is S-retractible over P_k. Suppose, in the first place, that $n > 2k$. Then $S(nL)$ is retractible, by suspension theory, and a retraction $f : S(nL) \to S_n$ determines a map $g : S_n \times S_k \to S_n$ such that

$$g(u, v) = g(uz, vz) \quad (u \in S_n, \ v \in S_k, \ z \in S).$$

The adjoint $g' : S_n \to E_{n,k}$ of g is a cross-section, since f is a retraction, and g' can be deformed into a cross-section of $O_{n,k}$, by

(7.3), since $n > 2k$.

Finally, suppose that $n \equiv 0 \bmod t_k$ with $n \leq 2k$. Then $P_{m,k}$ and $P_{m+n,k}$ have the same S-type, for all $m \geq k$. In the real case it follows from (1.6) and (1.7) that $k \leq 9$ and hence that $t_k = a_k$, the Hurwitz-Radon number. The complex case when $k = 2$ has already been dealt with at the end of §5. In the complex case when $k \geq 3$ it follows from the formulae (3.1), (3.2) for the cohomology operations that n is divisible by $3k$ or $4k$ according as k is or is not a power of 2, hence $n > 2k$ in either case. The quaternionic case is similar and so the proof of (7.2) is complete.

Let us now look at this result from the homotopy-equivariant point of view. Take the real case, first of all, and regard $V_{n,k}$ as a Z_2-space by changing the sign of the last column, as in §1. For $V_{n,1}$, in particular, the degree of the action is $(-1)^n$. The fibration $V_{n,k} \to V_{n,1}$ is equivariant and we can ask, as in §1, when there exist homotopy-equivariant cross-sections. For this we regard P^{k-1} as a Z_2-space, using reflection in the hyperplane of the last coordinate, and the Hopf line bundle as a Z_2-vector bundle similarly, just as in §6. Returning to the beginning of this section we find that a homotopy-equivariant cross-section $f : V_{n,1} \to V_{n,k}$ determines a fibre homotopy-equivariant equivalence

$$f'' : S(n\underline{R}) \to S(nL),$$

over P^{k-1}. This leads to

Theorem (7.4). Let $\hat{a}_k$ be the J/Z_2-order of the Hopf line bundle over P^{k-1}. Then $V_{n,k}$ admits a homotopy-equivariant cross-section if and only if $n \equiv 0 \bmod \hat{a}_k$.

The proof is very similar to that of (7.2). The only difficulty occurs in the 'if' part when $n \leq 2k$. Using (7.2), however, we see that in this range either $V_{n,k}$ admits a cross-section with k odd, or $V_{n,k+1}$ admits a cross-section. Thus in either case $V_{n,k}$ admits a homotopy-equivariant cross-section, as required.

The complex Stiefel manifolds provide another example. Consider the self-map T of $W_{n,k}$ given by complex conjugation. The degree of the action on $W_{n,1}$ is $(-1)^n$. The fibration $W_{n,k} \to W_{n,1}$ is equivariant, with respect to this action of Z_2, and again we can ask whether there exist homotopy-equivariant cross-sections. For example the cross-section

$$Sp(1) = X_{1,1} \subset W_{2,2} = U(2)$$

is homotopy-equivariant, since T transforms $z \in S(H)$ into jzj^{-1}. To discuss this question we regard $P_k(C)$ as a Z_2-space, under complex conjugation, and $L = L_C$ as a Z_2-vector bundle similarly. Then the argument used to prove (7.2) yields

Theorem (7.5). _Let_ $\hat{b}_k$ _be the_ J/Z_2_-order of the Hopf line bundle over_ $P_k(C)$. _Then_ $W_{n,k}$ _admits a homotopy-equivariant cross-section if and only if_ $n \equiv 0 \bmod \hat{b}_k$.

A further result of the same type will be given in the next section. The numbers $\hat{a}_k$ and $\hat{b}_k$ will be calculated later.

Returning to the general case we now introduce a second form of the intrinsic map which enables some further relationships to be established, as follows. Points of the cone CX on a space X are written in the usual form tx, where $t \in I$ and $x \in X$. Let M be an S-module so that $V = (M \times S_k)/S$ is a vector bundle over P_k. Consider the map

$$\theta : C(O_{n,k}) \times M \times S_k \to M \times A^n \times S_k$$

which is given by

$$(7.6) \quad \theta(tx,\ y,\ z) = ((1 - t^2)^{\frac{1}{2}}y,\ tx(z),\ z).$$

Clearly θ induces a map

$$(C(O_{n,k}) \times B(V),\ C(O_{n,k}) \times S(V) \cup O_{n,k} \times B(V)) \to (B(V \oplus nL),\ S(V \oplus nL))$$

and hence a map

$$(C(O_{n,k})/O_{n,k}) \wedge (B(V)/S(V)) \to B(V \oplus nL)/S(V \oplus nL).$$

In other words θ determines a map

$$\phi : S(O_{n,k}) \wedge P_k^V \to P_k^{V \oplus nL} .$$

We refer to this as the <u>intrinsic map of the second type.</u>

Now suppose that $O_{n,k}$ admits a cross-section $f : S_n \to O_{n,k}$. We already know, from (7.1) and the results of §5, that P_k^V and $P_k^{V\oplus nL}$ have the same S-type, but now we can make this more precise by proving

Proposition (7.7). <u>The composition</u>

$$f' = \phi(Sf \wedge 1) : S(S_n) \wedge P_k^V \to P_k^{V\oplus nL}$$

<u>is a homotopy equivalence.</u>

The proof proceeds by induction on k. When $k = 1$ the domain and codomain are spheres and it is easy to see that ϕ has degree ±1. Let $k \geq 2$, and suppose that (7.7) is true with $k - 1$ in place of k, $W = V|P_{k-1}$ in place of V, and $g = pf : S_n \to O_{n,k-1}$ in place of f. Now f' determines a homomorphism, raising dimension by dn, of the homology exact sequence of the pair (P_k^V, P_{k-1}^W) into that of the pair $(P_k^{V\oplus nL}, P_{k-1}^{W\oplus nL})$. This is an isomorphism of the relative groups, by another degree argument, and f'_* agrees with g'_* on the subspace. Hence the inductive step follows using the five lemma. Full details for the case when V is a multiple of L are given in [69]. In a sense (7.7) is a homological counterpart of (2.7), the generalized Freudenthal theorem. A very similar argument leads to

Proposition (7.8). <u>Suppose that there exists a retraction</u>

$$\rho : P_k^{V\oplus nL} \to S_{m+n},$$

<u>where</u> $m = \dim V$. <u>Then</u>

$$\rho\phi : S(O_{n,k}) \wedge P_k^V \to S_{m+n}$$

<u>restricts to a duality map</u>

$$S(Q_{n,k}) \wedge P_k^V \to S_{m+n}.$$

Corollary (7.9). _If_ $m + n \equiv k \bmod t_k$ _then_ $P_{m,k}$ _and_ $Q_{n,k}$ _are S-dual._

The latter also follows from Atiyah's duality theorem [7] for Thom spaces of vector bundles over a manifold. However, the rather specific form of duality map given here enables us to prove

Proposition (7.10). _If_ $m \equiv 0 \bmod t_k$ _and_ $m \geq 2k$ _then_ $S^{dm}Q_{n,k}$ _is a retract of_ $S^{dm}O_{n,k}$.

Corollary (7.11). _If_ $n \geq k$ _then_ $Q_{n,k}$ _is an S-retract of_ $O_{n,k}$.

We take $V = (m - k)L \oplus L'$ so that

$$\phi : S(O_{n,k}) \wedge Q_{m,k} \to Q_{m+n,k}.$$

The hypothesis ensures the existence of a section $g : S_m \to Q_{m,k}$, and hence a map

$$g' = \phi(1 \wedge g) : S(O_{n,k}) \wedge S_m \to Q_{m+n,k}.$$

Now an argument similar to the one used for proving (7.7) (see [69] for details) shows that

$$g'' : S(Q_{n,k}) \wedge S_m \to Q_{m+n,k}$$

is a homotopy-equivalence, where g'' denotes the restriction of g'. By composing g' with a homotopy inverse

$$Q_{m+n,k} \to S(Q_{n,k}) \wedge S_m$$

of g'' we obtain a retraction

$$S(O_{n,k}) \wedge S_m \to S(Q_{n,k}) \wedge S_m,$$

as required. Thus there exists a number r, depending on k but not on n, such that $S^rQ_{n,k}$ is a retract of $S^rO_{n,k}$. It would be interesting to know something about the least number r_k with this property. For example, is r_k bounded? And how is r_{k+1} related to r_k? It is easy to see that $r_k > 0$ when $k \geq 2$, also $r_2 > d$.

It seems likely that there is some relation between our map

$$\phi : S(O_{n,k}) \wedge Q_{m,k} \to Q_{m+n,k}$$

and the intrinsic map

$$h : O_{n,k} * O_{m,k} \to O_{m+n,k}$$

of §2, but so far as I know this has not been investigated.

These relationships can also be discussed from the homotopy-equivariant point of view, as follows. We take the real case. Consider the self-map d_r $(r = 1, \ldots, n)$ of $P_{n,k}$ given by reflection in the hyperplane of the rth coordinate. Clearly $d_1, \ldots, d_{n-k}$ belong to the same homotopy class α, say, while $d_{n-k+1}, \ldots, d_n$ belong to the same homotopy class β, say. Also $\alpha^{n-k} = \beta^k$, since changing the sign of all the coordinates gives the identity self-map. It is clear from the general discussion in §6 that if $P_{n,k}$ is regarded as a Z_2-space with action of type α then the stable homotopy-Z_2 type depends on the residue class of n mod a_k, while if $P_{n,k}$ is regarded as a Z_2-space with action of type β then the stable homotopy-Z_2 type depends on the residue class of n mod $\hat{a}_k$. To discuss the duality law we return to (7.8) with $m + n \equiv k \bmod \hat{a}_k$, and choose ρ to be a homotopy-equivariant retraction of the Thom space of $(m + n - k)L$ over P^{k-1} with Z_2-structure as before. We deduce that $P_{m,k}$ with β-structure is S-dual to $P_{n,k}$ with $\alpha\beta$-structure. In this sense the dual of β is $\alpha\beta$, hence the dual of $\alpha\beta$ is β, hence the dual of $\alpha = (\alpha\beta)\beta$ is $\beta(\alpha\beta) = \alpha$. Other ways to obtain these relations are given in [82] and [85].

Finally, let us regard $P_{n,k}$ as a subspace of $V_{n,k}$. For $r = 1, \ldots, n - k$ the self-map d_r of $P_{n,k}$ extends to the self-map of $V_{n,k}$ which changes the sign of the rth row, while for $r = n-k+1, \ldots, n$ it extends to the self-map of $V_{n,k}$ which changes the sign of the rth row and $(r - n + k)^{th}$ column. If we regard $(V_{n,k}, P_{n,k})$ as a Z_2-pair, in one of these ways, then the argument used to prove (7.11) shows that $P_{n,k}$ is a homotopy-Z_2 retract of $V_{n,k}$.

8·Relative Stiefel manifolds

This section is based on a recent article [83]. By the relative Stiefel manifolds I mean the pairs of spaces

$$W'_{n,k} = (V_{2n,2k}, W_{n,k}), \quad X'_{n,k} = (W_{2n,2k}, X_{n,k}).$$

In what follows I will concentrate on the former, leaving the corresponding results for the latter as exercises.

There is a notable formal analogy between the relative homotopy groups of $W'_{n,k}$ and the absolute homotopy groups of $W_{n,k}$. To see this, consider the factor space $\Gamma_n = O_{2n}/U_n$ $(n = 1, 2, \ldots)$, with the obvious embeddings $\Gamma_1 \subset \Gamma_2 \subset \ldots$. The triad homotopy group

$$\pi_r(O_{2n}; U_n, O_{2n-2k})$$

can be identified with the relative homotopy group

$$\pi_r(O_{2n}/U_n, O_{2n-2k}/U_{n-k})$$

on the one hand, or with

$$\pi_r(O_{2n}/O_{2n-2k}, U_n/U_{n-k})$$

on the other. Thus we can identify

$$\pi_r(W'_{n,k}) = \pi_r(\Gamma_n, \Gamma_{n-k}).$$

If $1 \leq l < k$, therefore, the homotopy exact sequence of the triple $(\Gamma_n, \Gamma_{n-l}, \Gamma_{n-k})$ can be written in the form

$$\ldots \to \pi_r(W'_{n-l,k-l}) \xrightarrow{u'_*} \pi_r(W'_{n,k}) \xrightarrow{p'_*} \pi_r(W'_{n,l}) \to \ldots$$

where u' denotes the inclusion and p' the natural projection.

Note that the inclusion

$$(V_{2n-1,1}, e) \to (V_{2n,2}, W_{n,1})$$

induces an isomorphism

$$\pi_r(S^{2n-2}) \approx \pi_r(W'_{n,1}),$$

for all values of r. The image of the generator $\iota_{2n-2} \in \pi_{2n-2}(S^{2n-2})$ will be denoted by $\kappa_{2n-2} \in \pi_{2n-2}(W'_{n,1})$. By a relative cross-section of $W'_{n,k}$ we mean an element of $\pi_{2n-2}(W'_{n,k})$ (or the representative of such an element) which projects into κ_{2n-2} under

$$p'_* : \pi_{2n-2}(W'_{n,k}) \to \pi_{2n-2}(W'_{n,1}).$$

For example, take $n = k$. Then the relative Stiefel manifold $W'_{n,n} = (O_{2n}, U_n)$ admits a relative cross-section if and only if the fibration $\Gamma_n \to S^{2n-2}$ admits a cross-section in the ordinary sense, i.e. if and only if $n = 2$ or 4 (see [26]). It would be interesting to know, incidentally, whether there exist fibrations of Γ_n with fibre Γ_{n-k} for $k \geq 2$; presumably not.

We have already noted that $V_{2n,2k}$ admits a cross-section if $W_{n,k}$ does. It is also true that $V_{2n,2k}$ admits a cross-section if $W'_{n,k}$ admits a relative cross-section, thus generalizing the result of Kirchhoff [93] (the case $n = k$). One way to see this is to consider the family of maps

$$B : \Gamma_n \to \Omega R_{2n}$$

given for $x \in O_{2n}$ and $t \in I$ by the commutator

$$B(xU_n)(t) = [x, e \cos \pi t + b \sin \pi t],$$

where

$$b = \begin{pmatrix} 0 & 1 \\ -1 & 0 \end{pmatrix} \oplus \dots \oplus \begin{pmatrix} 0 & 1 \\ -1 & 0 \end{pmatrix} \qquad (n \text{ summands}).$$

The maps B are compatible, for various n, and so determine a homomorphism

$$B_* : \pi_r(\Gamma_n, \Gamma_{n-k}) \to \pi_{r+1}(R_{2n}, R_{2n-2k}),$$

for all values of r. A straightforward calculation (see [75] for details) shows that

$$B_* : \pi_{2n-2}(\Gamma_n, \Gamma_{n-1}) \to \pi_{2n-1}(R_{2n}, R_{2n-2})$$

maps κ_{2n-2} onto an element $B_*\kappa_{2n-2}$ such that $p_*B_*\kappa_{2n-2}$ generates $\pi_{2n-1}(R_{2n}, R_{2n-1})$. By naturality, therefore, B_* transforms relative cross-sections of $W'_{n,k}$ into cross-sections of $V_{2n, 2k}$.

It is easy to check that the diagram shown below is commutative, where the verticals are inclusions and the horizontals are intrinsic maps.

$$\begin{array}{ccc} W_{m,k} * W_{n,k} & \to & W_{m+n,k} \\ \downarrow & & \downarrow \\ V_{2m,2k} * V_{2n,2k} & \to & V_{2m+2n,2k} \end{array}$$

Hence the <u>relative intrinsic map</u>

$$h : W_{m,k} * W'_{n,k} \to W'_{m+n,k}$$

is defined, and hence the <u>relative intrinsic join.</u> This construction has properties analogous to those found in the ordinary case. For example, the relative intrinsic join of a cross-section of $W_{m,k}$ and a relative cross-section of $W'_{n,k}$ is a relative cross-section of $W'_{m+n,k}$. Hence and from the results of §5 we obtain

Proposition (8.1). *The relative Stiefel manifold $W'_{n,k}$ admits a relative cross-section if $k = 1$ or*

$k = 2$ and $n \equiv 0 \bmod 2$, or

$k = 3$ or 4 and $n \equiv 4 \bmod 24$.

In §10 below we shall prove

Theorem (8.2). *The relative Stiefel manifold $W'_{n,k}$ admits no relative cross-section unless n and k are as in (8.1).*

In particular, $W'_{n,k}$ admits no relative cross-section for $k > 4$. One of the main steps in the proof of (8.2) is

Theorem (8.3). _If the relative Stiefel manifold $W'_{n,k}$ admits a relative cross-section then the sphere-bundles $S((n-1)C \oplus (L \otimes L))$ and $S(nL)$ have the same fibre homotopy type over $P_k(C)$._

The spheres $S(R^{2m})$ $(m = 0, 1, \dots)$ obtain S^1-structure through $S(C^m)$, as before. Thus elements of $V_{2n,2k}$ determine S^0-maps of S^{2k-1} into S^{2n-1}, while elements of $W_{n,k} \subset V_{2n,2k}$ determine S^1-maps. Any map

$$s : (B^{2n-2}, S^{2n-3}) \to (V_{2n,2k}, W_{n,k})$$

determines a map

$$s' : B^{2n-2} \times S^1 \times S^{2k-1} \cup S^{2n-3} \times B^2 \times S^{2k-1} \to S^{2n-1} \times S^{2k-1}$$

as follows. Since $s(x) = s_x$ is an S^0-map for all $x \in B^{2n-2}$ we define

$$s'(x, u^2, v) = (us_x(u^{-1}v), v) \quad (u \in S^1, v \in S^{2k-1}).$$

If $x \in S^{2n-3}$ then s_x is an S^1-map and so $s'(x, u^2, v) = (s_x v, v)$. We therefore complete the definition of s' by defining

$$s'(x, u^2, v) = (s_x v, v) \quad (x \in S^{2n-3}, u \in B^2, v \in S^{2k-1}).$$

Notice that if $s'(x, u^2, v) = (y, v)$, where $y \in S^{2n-1}$, then $s'(x, u^2z^2, vz) = (yz, vz)$, for any $z \in S^1$. Hence s' determines a fibre-preserving map

$$h : S((n-1)C \oplus (L \otimes L)) \to S(nL).$$

I assert that h is a fibre homotopy equivalence if the original map s is a relative cross-section.

Perhaps the easiest way to see this is to pull the situation back to real projective space through the standard fibration

$$f : P(R^{2k}) \to P(C^k).$$

Let L_C denote the canonical complex line bundle over $P(C^k)$ and L_R the canonical real line bundle over $P(R^{2k})$. The real bundle underlying

f^*L_C is $2L_R$, while the real bundle underlying $f^*(L_C \otimes L_C)$ is $2\underline{R}$. Thus h pulls back to a fibre-preserving map

$$h' : S(2nR) \rightarrow S(2nL),$$

over $P(R^{2k})$. The degree of h' over the point-space $P(R^1)$ is easily computed to be ±1, as in the proof of (7.1). Hence h has degree ±1 on the fibres and (8.3) follows from Dold's theorem. In fact h' can be identified with the fibre-preserving map corresponding to s', the cross-section of $V_{2n, 2k}$ determined by s as described above. An alternative proof of (8.3) can be found in [83].

It turns out, as we shall see, that the converse of (8.3) is also true, although it does not seem easy to establish this by either of the methods used in §7.

9·Cannibalistic characteristic classes

In this section we shall complete the proof of (1. 2), following Bott [31] and Adams [3], and go on to complete the proof of (1. 10) by the same method. We begin by summarizing the relevant theory.

Let V be a euclidean bundle over a space X. If V is J-trivial then the Thom space X^V is S-retractible, as we have seen. Hence by computing the Steenrod squares in the cohomology of X^V we can expect to obtain some information about the J-order of V. A more direct approach, however, is to calculate the Stiefel-Whitney classes of V in the cohomology of X, since these must vanish if V is J-trivial. The link between these two approaches is given by the theorem of Thom [140], which expresses the Stiefel Whitney classes of V in terms of the action of the Steenrod operations in the cohomology of X^V, using the Thom isomorphism

$$H^m(X) \approx \tilde{H}^{m+n}(X^V) \qquad (n = \dim V).$$

In K-theory the position is similar. By computing the Adams ψ-operations in $\tilde{K}_R(X^V)$ we can obtain some information about the J-order of V, but there is also a direct approach which involves calculating certain characteristic classes of V, with values in $K_R(X)$. To define these cannibalistic classes, as they are called, we need the Thom isomorphism theorem of real K-theory and for this certain restrictions on V are necessary.

Recall that V admits spin-structure if the Stiefel-Whitney classes $w_1(V)$ and $w_2(V)$ both vanish. Let us say that V is admissible, in the present context, if V admits spin structure and $n = \dim V$ is a multiple of 8.

Since $X^V = B(V)/S(V)$, by definition, we can identify $\tilde{K}_R(X^V)$ with $K_R(B(V), S(V))$. Thus $\tilde{K}_R(X^V)$ becomes a module over $K_R(B(V))$

and hence over $K_R(X)$, through the homomorphism p^* induced by $p : B(V) \to X$. If V is admissible then the Thom isomorphism

$$\phi = \phi_V : K_R(X) \to \tilde{K}_R(X^V)$$

is defined, as in [32], and is an isomorphism of modules. Moreover the 'cannibalistic' characteristic class $\rho^t(V) \in K_R(X)$ is defined by

$$(9.1) \quad \rho^t(V) = \phi^{-1}\psi^t\phi(1) \qquad (t \in Z).$$

Since ϕ reduces to the periodicity isomorphism when V is trivial it follows that

$$(9.2) \quad \rho^t(\underline{n}) = t^{n/2} \qquad (n \equiv 0 \bmod 8).$$

Following Adams in Part II of [4] we now prove

Theorem (9.3). Let U, V be admissible bundles over X such that $S(U)$, $S(V)$ have the same fibre homotopy type. Then

$$\rho^t(U) \,.\, \psi^t(1 + x) = \rho^t(V) \,.\, (1 + x)$$

for some element $x \in \tilde{K}_R(X)$.

The argument is similar to Thom's proof [140] that Stiefel-Whitney classes are invariants of the J-type. To simplify notation we will omit the index t. Given a map $h : X^U \to X^V$ I assert that

$$(9.4) \quad y \,.\, \rho(V) = \rho(U) \,.\, \psi(y),$$

where $y = \phi_u^{-1}h^*\phi_V(1) \in K_R(X)$. For since h^* commutes with ψ we have at once

$$(\phi_U^{-1}h^*\phi_V)(\phi_V^{-1}\psi\phi_V) = (\phi_U^{-1}\psi\phi_U)(\phi_U^{-1}h^*\phi_V).$$

Evaluating at the identity of $K_R(X)$ this yields

$$(\phi_U^{-1}h^*\phi_V) \,.\, \rho(V) = (\phi_U^{-1}\psi\phi_U) \,.\, y.$$

However since ϕ_U, ϕ_V and h^* are morphisms of modules this can be rewritten

$$(\phi_U^{-1} h^* \phi_V) 1 \,.\, \rho(V) = \phi_U^{-1} \psi(\phi_U(1) \,.\, y)$$
$$= \phi_U^{-1} \psi \phi_U(1) \,.\, \psi(y),$$

since ψ is a ring homomorphism, and so (9. 4) is obtained.

Now suppose that h is given by a fibre-preserving map $S(U) \to S(V)$ of degree d, say, on the fibres. Then $b^*(y) = d$, where

$$b^* : K_R(X) \to K_R(pt) = Z,$$

is induced by the inclusion of the basepoint. Thus $y - d$ lies in the kernel $\tilde{K}_R(X)$ of b^*. When $d = \pm 1$ we multiply (9. 4) by d and obtain (9. 3) with $x = yd - 1$. Note that $1 + x$ is invertible, either for algebraic reasons or from consideration of the inverse fibre homotopy equivalence. Thus (9. 2) and (9. 3) imply

Corollary (9. 5). _If_ $S(U)$ _is trivial, in the sense of fibre homotopy type then_

$$\rho^t U = t^{n/2} \frac{\psi^t(1 + x)}{1 + x},$$

for some $x \in \tilde{K}_R(X)$, _where_ $n = \dim U$.

It turns out that ρ^t has the property that

(9. 6) $\rho^t(U \oplus V) = (\rho^t U) \,.\, (\rho^t V),$

for all admissible line bundles U, V. Moreover a straightforward calculation using representation theory (see [4]) shows that

(9. 7) $\rho^t(nL) = t^{n/2} + \frac{1}{2} t^{n/2}([L] - 1)$ (t even)

$= t^{n/2} + \frac{1}{2}(t^{n/2} - 1)([L] - 1)$ (t odd),

for any line bundle L and $n \equiv 0 \bmod 8$.

For example, take $X = P^{k-1}$ with L the Hopf line bundle. Recall that $\tilde{K}_R(P^{k-1})$ is cyclic of order a_k and that $\alpha = [L] - 1$ is a generator. We are now ready to prove that

(9.8) $J : \tilde{K}_R(P^{k-1}) \approx J(P^{k-1})$.

For this it is sufficient to show that the order of $J(\alpha)$ is a_k. The first step is to use Stiefel-Whitney classes. Since

$$w_m(nL) = \binom{n}{m} a^m \quad (m = 1, 2, \dots),$$

where a generates $H^1(P^{k-1})$, this shows (cf. (1.6)) that the J-order of L is divisible by 2^r, where r is the least integer such that $2^r \geq k$. This proves (9.8) for $k \leq 4$ and shows, for $k \geq 5$, that the J-order of α is a multiple of 8. Let $k \geq 5$, therefore, and suppose that nL is J-trivial. Then $n \equiv 0 \bmod 8$ and so (9.7) applies. Let t be odd, so that ψ^t acts trivially on $\tilde{K}_R(P^{k-1})$. After increasing n by multiples of a_k, if necessary, the sphere-bundle $S(nL)$ will be trivial, in the sense of fibre homotopy type, and so

$$\tfrac{1}{2}(t^{n/2} - 1)\alpha = 0,$$

by (9.5) and (9.7). Take $t = 3$, in particular, and use (1.9) as before. We conclude that $n/2$ is an even multiple of $2^{\sigma(k)-2}$, hence n is a multiple of $2^{\sigma(k)} = a_k$. This completes the proof. Finally we obtain (1.2), Adams' theorem, by combining (7.2) and (9.8).

Turning to the homotopy-equivariant problem, recall that $\hat{P}^{k-1}$ is constructed from P^{k-1} with respect to reflection in the hyperplane of of the last coordinate x_k. Let $\hat{P}^k$ be constructed from P^k with respect to reflection in the hyperplane of the same coordinate x_k. Then $\hat{P}^{k-1} \subset \hat{P}^k$ with inclusion u, say, and we have an exact sequence

$$\tilde{K}_R(\hat{P}^k) \xrightarrow{u^*} \tilde{K}_R(\hat{P}^{k-1}) \xrightarrow{\delta} \tilde{K}^1_R(S(\hat{P}^k/\hat{P}^{k-1})).$$

Let $k \equiv 0 \bmod 4$. Then the group on the right is cyclic infinite, since $\hat{P}^k/\hat{P}^{k-1}$ has the homotopy type of $S^{k+1} \vee S^k$, while the group in the centre is finite. Therefore u^* is surjective. Now the action on P^k is neutral, since k is even, and so $\hat{P}^k$ has the same homotopy type as $P^k \times S^1$. Hence it follows, after a little calculation, that ψ^t (t odd) operates trivially on $\tilde{K}_R(\hat{P}^k)$ and hence on $\tilde{K}_R(\hat{P}^{k-1})$. Now apply (9.5) and (9.7), just as before. We conclude that the J-order of $\hat{L}$ over $\hat{P}_k$

is a multiple of $a_{k+1} = 2a_k$. Taken with our previous results, this proves

Proposition (9.9). *The* J/Z_2*-order* $\hat{a}_k$ *of the* Z_2*-line bundle* L *over* P^{k-1} *is given by* $\hat{a}_k = a_{k+1} = 2a_k$ *when* $k = 2$ *or* $k \equiv 0 \bmod 4$, *by* $\hat{a}_k = a_k$ *otherwise.*

Hence and from (7.4), (1.10) follows at once.

10 · Exponential characteristic classes

In this section and the next we shall be studying the problem of the existence of cross-sections for complex and quaternionic Stiefel manifolds. As we have seen, this depends on a knowledge of the J-order of the canonical line bundle over the appropriate projective space. Although in principle this can be determined by means of the cannibalistic classes of §9, the calculations involved are formidable. It turns out to be far more convenient to use what Adams calls the Bernoulli class bh in the complex case, and the hyperbolic class sh in the quaternionic. The applications will be given after we have discussed the basic properties of these two exponential characteristic classes.

In what follows we shall be working with both real K-theory K_R and complex K-theory K_C. The homomorphisms

$$K_C(X) \xrightarrow{R_\#} K_R(X) \xrightarrow{C_\#} K_C(X) \xrightarrow{T_\#} K_C(X)$$

are important, where $R_\#$ is given by the underlying real bundle, $C_\#$ means tensoring with C, and $T_\#$ denotes complex conjugation. Recall that

$$(10.1) \quad R_\# C_\# = 2, \quad C_\# R_\# = 1 + T_\#, \quad T_\# C_\# = C_\#.$$

We shall need a few basic facts about the K-theory of projective spaces, which can be found in §2 of [5] and §3 of [123]. First, the Chern character shows that

$$(10.2) \quad \tilde{K}_C(P_k(C)) = Z[\beta] \bmod \beta^k,$$

where $\beta = [L] - 1$ and $L = L_C$ is the Hopf line bundle. Next, a spectral sequence argument (for example) shows that $R_\# \beta$ generates $\tilde{K}_R(P_k(C))$, and so

(10.3) $R_\# \tilde{K}_C(P_k(C)) = \tilde{K}_R(P_k(C))$.

More detailed calculations, as in [5], show that the sequence

(10.4) $\tilde{K}_R(P_k(C)) \xrightarrow{C_\#} \tilde{K}_C(P_k(C)) \xrightarrow{1-T_\#} \tilde{K}_C(P_k(C))$

is exact. Turning to the quaternionic case, let L^C denote the complex 2-plane bundle underlying the Hopf bundle $L = L_H$ over $P_k(H)$. The Chern character shows that

(10.5) $\tilde{K}_C(P_k(H)) = Z[\gamma] \bmod \gamma^k$,

where $\gamma = [L^C] - 2$, and moreover

(10.6) $g^*\gamma = (1 + T_\#)\beta$,

where $g : P_{2k}(C) \to P_k(H)$ denotes the standard fibration. Finally, we have

Proposition (10.7). _The homomorphism_

$C_\# : \tilde{K}_R(P_k(H)) \to \tilde{K}_C(P_k(H))$

is injective, with image the subring generated by 2γ _and_ γ^2.

For any finite complex X the Adams _Bernoulli_ characteristic class is a homomorphism

$$bh : K_C(X) \to 1 + \sum_{s>0} H^{2s}(X; Q)$$

from the _additive_ Grothendieck group of complex vector bundles to the _multiplicative_ group of rational cohomology in even dimensions with augmentation unity. Such a homomorphism is called _exponential_ and is determined by its value for line bundles. In the case of bh the value is

(10.8) $bh(L) = (e^y - 1)/y$,

where $y = c_1(L)$ is the first rational Chern class of any line bundle L.

Alternatively bh can be defined by taking the Chern character in the Thom space. To be precise, let V be a complex vector bundle over X. Then

$$(10.9) \quad bh(V) = \phi^{-1} ch \phi(1),$$

as shown in the following diagram where ϕ denotes the Thom isomorphism for both complex K-theory and rational cohomology.

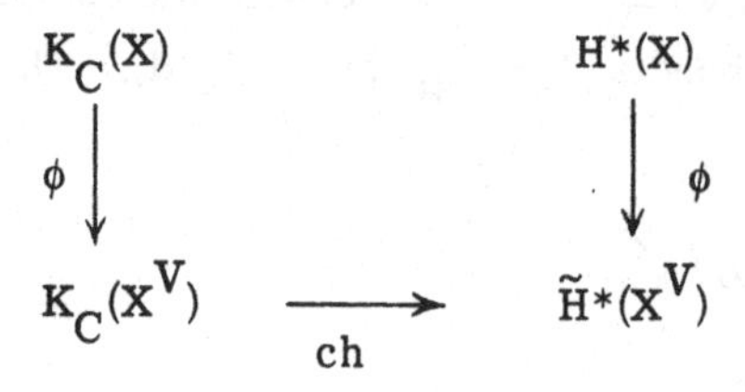

The equivalence of these two definitions follows from (14.3) of [9]. Using the latter version we can apply the Thom invariance argument, as in the proof of (9.5), and obtain

Theorem (10.10). _If_ $\xi \in K_C(X)$ _is_ J-_trivial then_

$$bh(\xi) \in ch(1 + \tilde{K}_C(X)).$$

Following Adams and Walker [5], let us say that ξ is J'_C-_trivial_ if the condition in (10.10) is satisfied, and define $J'_C(X)$ to be the factor group of $\tilde{K}_C(X)$ by the subgroup of J'_C-trivial elements.

For example, take $X = P_k(C)$, where $k \geq 2$. The integral cohomology ring is given by

$$H^*(P_k(C)) = Z[b] \bmod b^k,$$

where $b = c_1(L)$ is the first Chern class of the Hopf line bundle L. Regarding b as a rational class we see at once that $ch\beta^j \equiv b^j$, modulo higher powers of b, for $j = 0, 1, \ldots$. Thus any element of $H^*(P_k(C); Q)$ has a unique expression of the form

$$p_0 ch\beta^0 + p_1 ch\beta^1 + \ldots + p_{k-1} ch\beta^{k-1},$$

with rational coefficients $p_0, p_1, \ldots, p_{k-1}$. Moreover the element lies in the image of

$$\mathrm{ch} : 1 + \tilde{K}_C(P_k(C)) \to H^*(P_k(C); Q)$$

if and only if all these coefficients are integers. Write $y = \mathrm{ch}\beta = e^b - 1$, so that $b = \log(1 + y)$, defined by its usual power series expansion. Then

$$bh(-nL) = \left(\frac{e^b - 1}{b}\right)^{-n} = \left(\frac{\log(1 + y)}{y}\right)^n,$$

for any $n \geq 0$, and so we obtain

Theorem (10. 11). The element $n\beta \in \tilde{K}_C(P_k(C))$ is J'_C-trivial if and only if the coefficients of $y^1, \ldots, y^{k-1}$ are integers in the expansion of

$$\left(\frac{\log(1 + y)}{y}\right)^n.$$

Combining this with (10. 10) we obtain a necessary condition for $n\beta$ to be J-trivial and hence for $W_{n,k}$ to have a cross-section, by (7. 2). In the next section we shall show, following Adams and Walker, that the necessary condition is also sufficient. It was Atiyah and Todd [11] who first showed the condition to be necessary, using a straightforward Chern character argument, and they went on to translate the condition into another form as follows. For any prime p and integer x let $\nu_p(x)$ denote the exponent of the highest power of p which divides x. Let b'_k $(k = 1, 2, \ldots)$ denote the J'_C-order of $\beta \in \tilde{K}_C(P_k(C))$, as determined by (10. 11). Atiyah and Todd show that b'_k is explicitly given by

$$(10.12) \quad \nu_p(b'_k) = \sup_r(r + \nu_p(r)) \quad \left(1 \leq r \leq \left[\frac{k - 1}{p - 1}\right]\right).$$

Thus $b'_1 = 1$, $b'_2 = 2$, $b'_3 = b'_4 = 24$, and so forth. Furthermore

$$(10.13) \quad b'_{k+1} = b'_k \quad (k \text{ odd}, k > 1).$$

For if not then $m' + \nu_p(m') > m + \nu_p(m)$, for some prime number p such that

$$m' = \left[\frac{k}{p - 1}\right] > m = \left[\frac{k - 1}{p - 1}\right].$$

The latter condition requires k to be a multiple of $p - 1$ and hence $p = 2$, since k is odd. But then

$$m' = k, \; m = k - 1, \; \nu_p(m') = 0, \; \nu_p(m) = 1,$$

and the former condition is violated.

Another application of the Adams-Bernoulli class is provided by

Theorem (10.14). The complex vector bundles $L \otimes L$ and nL over $P_k(C)$ are J'_C-equivalent if and only if $k = 1$ or $k = 2$ and $n \equiv 0 \bmod 2$, or $k = 3$ or 4 and $n \equiv 4 \bmod 24$.

By (10.10) these are necessary conditions for the bundles to be J-equivalent, and hence for the relative Stiefel manifold $W'_{n,k}$ to have a cross-section, by (8.3). Thus (10.14) implies (8.2).

To prove (10.14), observe first that $bh(L \otimes L) = (e^{2b} - 1)/2b$, by (10.8). Hence

$$\begin{aligned} bh[L \otimes L - nL] &= \left(\frac{e^{2b} - 1}{2b}\right)\left(\frac{e^{b} - 1}{b}\right)^{-n} \\ &= \left(1 + \frac{y}{2}\right)\left(\frac{\log(1 + y)}{y}\right)^{n-1}, \end{aligned}$$

by the exponential property, where $y = e^b - 1$. Thus $L \otimes L$ and nL are J'_C-equivalent if and only if the coefficients of $y^1, \ldots, y^{k-1}$ are integers, in the expansion. The coefficient of y^1 is $1 - n/2$; thus n is even if $k \geq 2$. The coefficient of y^2 is $(n - 1)(3n - 4)/24$; thus $n \equiv 4 \bmod 24$ if $k \geq 3$. The coefficient of y^3 is $(1 - n)(n^2 + 6n - 8)/48$; thus $n \equiv 4 \bmod 24$ if $k \geq 4$. Finally, the coefficient of y^4 is the quotient of

$$r(n) = 15n^4 + 30n^3 + 5n^2 - 18n - 32$$

by $5.9.128$. If $n \equiv 4 \bmod 24$ then since $n^2 \equiv 0 \bmod 16$ we have $r(n) \equiv 8 \bmod 16$, so that the quotient cannot be integral. Thus $k \leq 4$ and the proof of (10.14) is complete.

In these examples it turns out that the J-order and the J'_C-order are equal, but in general this will not be so; quaternionic projective space is a case in point. To deal with this Adams has introduced another exponential class, as follows.

The Adams <u>hyperbolic</u> class is a natural homomorphism

$$\mathrm{sh} : K_R(X) \to 1 + \sum_{s>0} H^{4s}(X; Q),$$

from the additive Grothendieck group of real vector bundles to the multiplicative group of rational cohomology with augmentation unity in dimensions divisible by 4. The homomorphism is exponential with defining property

$$(10.15) \quad \mathrm{sh}(R_{\#}L) = (e^{\frac{1}{2}y} - e^{-\frac{1}{2}y})/y,$$

where y denotes the first rational Chern class of L. The connection between bh and sh is given by

$$(10.16) \quad \mathrm{bh}(\xi) = e^{\frac{1}{2}c_1(\xi)}\mathrm{sh}(R_{\#}\xi)$$

where $\xi \in K_C(X)$. This is clear enough for line bundles and hence is true in general by the theory of exponential classes.

Just as in the case of bh there is an alternative definition in terms of the appropriate Thom isomorphism which implies

Theorem (10.17). <u>If</u> $\xi \in K_R(X)$ <u>is</u> J-<u>trivial then</u>

$$\mathrm{sh}(\xi) \in \mathrm{ch}C_{\#}(1 + \tilde{K}_R(X)).$$

Let us say that ξ is J'_R-<u>trivial</u> if $\mathrm{sh}(\xi)$ satisfies the condition in (10.17) and define $J'_R(X)$ to be the factor group of $K_R(X)$ by the subgroup of J'_R-trivial elements. Of course $J'_R(X)$ can also be regarded as a factor group of $J(X)$.

Let $\xi \in K_C(X)$ be an element such that $R_{\#}\xi$ is J'_R-trivial. In general ξ is not J'_C-trivial. Suppose, however, that $c_1(\xi) = 2h$ for some $h \in H^2(X)$. Then $h = c_1(\eta)$, for some complex line bundle η, and then $\mathrm{ch}(\eta) = \exp(\frac{1}{2}c_1(\xi))$. Thus if

$$\mathrm{sh}R_{\#}\xi = \mathrm{ch}C_{\#}(1 + \theta),$$

for some $\theta \in \tilde{K}_R(X)$, then

$$bh(\xi) = ch(\eta)ch(1 + C_\#\theta)$$
$$= ch\eta(1 + C_\#\theta) = ch(1 + \theta'),$$

where $\theta' = -1 + \eta(1 + C_\#\theta) \in \tilde{K}_C(X)$. Therefore ξ is J'_C-trivial if $R_\#\xi$ is J'_R-trivial and $c_1(\xi) \in 2H^2(X)$.

For example, take $X = P_k(C)$. I assert that $R_\#\xi$ is J'_R-trivial if and only if ξ is J'_C-trivial, for all $\xi \in K_C(P_k(C))$. To prove 'only if' we use the above argument; a short calculation shows that the further condition is automatically satisfied when $R_\#\xi$ is J'_R-trivial. Conversely, suppose that ξ is J'_C-trivial, so that

$$ch(1 + \theta) = bh(\xi) = \exp \tfrac{1}{2}c_1(\xi) sh\, R_\#\xi,$$

for some $\theta \in \tilde{K}_C(P_k(C))$. Comparing terms in dimension two we see that $\frac{1}{2}c_1(\xi) = c_1(\theta) = c_1(\phi)$, for some complex line bundle ϕ, and hence $sh(R_\#\xi) = ch(1 + \psi)$, as before, for some $\psi \in \tilde{K}_C(P_k(C))$. The proof is completed by showing that $T_\#\psi = \psi$ and so $\psi \in C\tilde{K}_R(P_k(C))$, by (10.4) above. To see this, consider the self-map T of $P_k(C)$ given by complex conjugation. Since $T^* = T_\#$ on the Hopf line bundle it follows that $T^* = T_\#$ on any element of $\tilde{K}_C(P_k(C))$. Thus

$$R_\#\xi = R_\#T_\#\xi = R_\#T^*\xi = T^*R_\#\xi,$$

and so

$$ch(1 + \psi) = T^*ch(1 + \psi) = ch(1 + T^*\psi) = ch(1 + T_\#\psi).$$

Therefore $\psi = T_\#\psi$, since the Chern character is injective, and the proof is complete. Combining this with (10.3) we see that

$$(10.18) \quad J'_R(P_k(C)) = J'_C(P_k(C)),$$

for all values of k. The same is true with P_k replaced by $P_{k,2}$ (k odd), by a very similar argument.

While retaining the J'_C notation, from now on we shall write J' instead of J'_R. This agrees with the usage of Adams and Walker [5], also Sigrist and Suter [130], but not the usage of Adams in [4].

For our last example we take $X = P_k(H)$, where $k \geq 2$. Recall that

$$H^*(P_k(H)) = Z[c] \bmod c^k,$$

where c generates $H^4(P_k(H))$. We choose c, as we may, so that c maps into b^2 under the homomorphism

$$g^* : H^*(P_k(H)) \to H^*(P_{2k}(C))$$

induced by the standard fibration. Now

$$g^* R_\# \gamma = R_\# g^* \gamma = R_\#(\beta + T_\# \beta) = 2R_\# \beta,$$

by (10. 1) and (10. 6). By definition

$$sh(R_\# \beta) = (e^{b/2} - e^{-b/2})/b = 2sh(b/2).$$

Hence and from (10. 7) it follows just as in the complex case that any element in the image of

$$g^* \circ ch \circ C_\# = ch \circ g^* \circ C_\# = K_R(P_k(H)) \to H^*(P_{2k}(C); Q)$$

has a unique expression of the form

$$\sum_{j=0}^{k-1} q_j e_j [2sh(b/2)]^{2j},$$

with rational coefficients $q_0, q_1, \ldots, q_{k-1}$, where $e_j = 1$ or 2 according as j is even or odd. Moreover the element lies in the image of

$$g^* \circ ch \circ C_\# : 1 + \tilde{K}_R(P_k(H)) \to H^*(P_{2k}(C)); Q)$$

if and only if all these coefficients are integers. Write $b = 2sh^{-1}(\sqrt{z}/2)$. Then

$$g^*\mathrm{sh}(-n.\, R_{\#}\beta) = \left(\frac{b}{2\mathrm{sh}(b/2)}\right)^{2n} = \frac{2}{\sqrt{z}}\, \mathrm{sh}^{-1}\left(\frac{\sqrt{z}}{2}\right)^{2n},$$

for any integer n, and we obtain

Theorem (10.19). The element $n\gamma \in \tilde{K}_C(P_k(H))$ is J'-trivial if and only if the coefficients of $z^1, \ldots, z^{k-1}$ in the expansion of

$$\frac{2}{\sqrt{z}}\, \mathrm{sh}^{-1}\left(\frac{\sqrt{z}}{2}\right)^{2n}$$

are integers or even integers according as the exponent of z is even or odd.

Combining this with (10.17) we obtain a necessary condition for $n\gamma$ to be J-trivial and hence for $X_{n,k}$ to have a cross-section, by (7.2). This result is due to Sigrist and Suter [130] who show, moreover, that the condition is sufficient, as we shall see in the next section. Sigrist and Suter go on to show that the J'_C-order c'_k of $\gamma \in \tilde{K}_C(P_k(H))$, determined by (10.19), is explicitly given by

$$(10.20) \qquad \begin{cases} \nu_2(c'_k) = \sup_r(2k-1,\ 2r + \nu_2(r)) \quad (1 \le r < k), \\ v_p(c'_k) = \nu_p(b'_{2k}) \quad (p \text{ odd}). \end{cases}$$

Comparing (10.12) and (10.20) we see that c'_k is equal to either b'_{2k} or $b'_{2k/2}$, for any given value of k. The latter alternative always holds when k is odd. It also holds for some of the even values of k, beginning with $k = 10$. An interesting statistical analysis of the situation is given at the end of [130].

11 · The main theorem of J-theory

The purpose of this section is to determine the J-order of the Hopf line bundle over complex and quaternionic projective space and then use (6.2) to deduce the cross-section theorems of Adams-Walker and Sigrist-Suter for the corresponding Stiefel manifolds. We require

Theorem (11.1). *For any* $x \in \tilde{K}_R(X)$ *and integer* t *there exists an integer* e *such that*

$$t^e(\psi^t - 1)x$$

is J-trivial.

This famous result was conjectured by Adams [4], who proved it in some special cases. The first proof of the theorem itself was given by Quillen [117], using methods inspired by algebraic geometry. Subsequently Becker and Gottlieb [21] have given a proof within the framework of conventional algebraic topology.

In view of (11.1) we follow Adams [4] by defining a functor $J''(X)$ as follows. Given a function f which assigns to each integer t a non-negative integer $f(t)$, let $W(f, X)$ denote the subgroup of $\tilde{K}_R(X)$ generated by the elements

$$t^{f(t)}(\psi^t - 1)x \qquad (x \in \tilde{K}_R(X),\ t \in Z).$$

Then we define $J''(X) = \tilde{K}_R(X)/W(X)$, where

$$W(X) = \bigcap_f W(f, X),$$

taken over all such functions f. Even without using (11.1) many useful results can be proved about $J''(X)$, as in Part II of [4]; for example $J''(X)$ is a finite group. However (11.1) shows that $J(X)$ can be regarded as a factor group of $J''(X)$. The main theorem of J-theory is

Theorem (11. 2). The natural projection

$$\theta'' : J''(X) \to J(X)$$

is an isomorphism.

This is a combination of (11. 1) and another result of Adams, proved in Part III of [4]. In principle, therefore, the calculation of the J-order of an element of $\tilde{K}_R(X)$ can be carried out, once the action of the ψ-operations is known. In practise, however, the calculations present formidable difficulties, even for such convenient spaces as the complex projective spaces. To get round this the procedure is to consider the composition of the natural projections

$$J''(X) \xrightarrow{\theta''} J(X) \xrightarrow{\theta'} J'(X).$$

Neither $J''(X)$ nor $J'(X)$ is particularly difficult to compute when X is a sphere or complex of a simple form. Let us say that X is a J'-space if $\theta' \circ \theta''$, and hence θ' is an isomorphism. Adams makes some calculations in Parts II and III of [4], which we shall omit, and shows that S^{4q} is a J'-space for $q \geq 1$. Furthermore the mapping cone $e^{4q} \cup S^{4q-2}$ of the generator η of $\pi_{4q-1}(S^{4q-2})$ is a J'-space for $q \geq 1$. These results will be used later.

Both J' and J'' have a limited exactness property, as shown by the following three results from [4] and [5]. Let X, Y, Z be finite complexes and let

$$X \xrightarrow{i} Y \xrightarrow{j} Z$$

be a cofibration, with inclusion i and projection j. We prove

Proposition (11. 3). Suppose that both

$$ch : K_C(X) \to H^*(X; Q), \quad j^* : H^*(Z; Q) \to H^*(Y; Q)$$

are injective. Then

$$j^* : J'_C(Z) \to J'_C(Y)$$

is also injective.

For let $\xi \in K_C(Z)$ be an element such that $bhj^*\xi = ch(1 + \theta)$, for some $\theta \in \tilde{K}_C(Y)$. Then

$$\begin{aligned} ch\, i^*(1 + \theta) = i^*ch(1 + \theta) &= i^*bh\, j^*\xi \\ &= bhi^*j^*\xi = 1, \end{aligned}$$

since $i^*j^* = 0$. Therefore $i^*(1 + \theta) = 1$ and so $1 + \theta = j^*(1 + \phi)$, by exactness, for some $\phi \in \tilde{K}_C(Z)$. Therefore

$$\begin{aligned} j^*bh(\xi) = bh(j^*\xi) &= ch(1 + \theta) \\ &= ch\, j^*(1 + \phi) = j^*ch(1 + \phi). \end{aligned}$$

Hence $bh(\xi) = ch(1 + \phi)$ and so ξ is J'_C-trivial, as asserted. A very similar argument proves

Proposition (11.4). Suppose that both

$$chC_{\#} : K_R(X) \to H^*(X; Q), \quad j^* : H^*(Z; Q) \to H^*(Y; Q)$$

are injective. Then

$$j^* : J'_R(Z) \to J'_R(Y)$$

is also injective.

Finally we turn to J'' and prove

Proposition (11.5). Suppose that the sequence

$$\tilde{K}_R(Z) \xrightarrow{j^*} \tilde{K}_R(Y) \xrightarrow{i^*} \tilde{K}_R(X) \to 0$$

is exact. Then the sequence

$$J''(Z) \xrightarrow{j^*} J''(Y) \xrightarrow{i^*} J''(X) \to 0$$

is also exact.

Recall that $J''(Z)$ is finite. Choose a finite set $(z_1, \ldots, z_q)$ of representatives in $\tilde{K}_R(Z)$ for the elements of $J''(Z)$. Suppose that $y \in \tilde{K}_R(Y)$ is an element such that $i^*y \in \bigcap_f W(f, X)$. I assert that for each such f there exists an element z_r such that $y - j^*z_r \in W(f, Y)$.

By hypothesis there exists a set of elements $x_t \in \tilde{K}_R(X)$, of which all but a finite number are zero, such that

$$i^*y = \sum_t t^{f(t)}(\psi^t - 1)x_t.$$

Since i^* is onto we can find $y_t \in \tilde{K}_R(Y)$ such that $x_t = i^*y_t$ and $y_t = 0$ whenever $x_t = 0$. Then

$$y - \sum_t t^{f(t)}(\psi^t - 1)y_t$$

lies in the kernel of i^* and so equals j^*z, by exactness, for some $z \in \tilde{K}_R(Z)$. However $J''z = J''z_r$, for some representative z_r. Since $z - z_r \in W(X) \subset W(f, X)$ we have

$$z - z_r = \sum_t t^{f(t)}(\psi^t - 1)z_t',$$

where $z_t' \in \tilde{K}_R(Z)$, and hence

$$y = j^*z_r + \sum_t t^{f(t)}(\psi^t - 1)(y_t + j^*z_t').$$

Thus $y - j^*z_r \in W(f, Y)$, as asserted.

Moreover there exists a representative z_r which satisfies this condition for all such functions f. For if not then for each representative z_r there exists a function f_r such that $y - j^*z_r \notin W(f_r, Y)$. Define a function f by

$$f(t) = \sup_{1 \leq r \leq q} f_r(t);$$

then for each r we have $y - j^*z_r \notin W(f, Y)$, contrary to what has already been established. We have shown, therefore, that

$$y - j^*z_r \in W(Y)$$

for some r, thus $J''y = J''j^*z_r = j^*J''z_r$. The rest of the proof of (11.5) is obvious.

Recall that X is a J'-space if

$$\theta = \theta' \circ \theta'' : J''(X) \to J'(X)$$

is an isomorphism. We prove

Lemma (11.6). *If* k *is odd then* $P_k(C)$ *is a* J'-*space.*

The result is trivial when $k = 1$. Let $k \geq 3$, therefore, and suppose that the result is true with $k - 2$ in place of k. Consider the cofibration

$$P_{k-2}(C) \xrightarrow{i} P_k(C) \xrightarrow{j} P_{k,2}(C),$$

which gives rise to the following commutative diagram.

$$\begin{array}{ccccccc}
J''(P_{k,2}(C)) & \xrightarrow{j^*} & J''(P_k(C)) & \xrightarrow{i^*} & J''(P_{k-2}(C)) & \rightarrow & 0 \\
\downarrow \theta & & \downarrow \theta & & \downarrow \theta & & \\
J'(P_{k,2}(C)) & \xrightarrow{j^*} & J'(P_k(C)) & \xrightarrow{i^*} & J'(P_{k-2}(C)) & &
\end{array}$$

The upper row is exact, by (11.5). Also j^* in the bottom row is injective, by (10.18) and (11.3). Also $P_{k,2}(C)$ is a J'-space, as noted above, and $P_{k-2}(C)$ is a J'-space, by hypothesis. Therefore $P_k(C)$ is a J'-space, and (11.6) follows by induction. The corresponding result for even values of k is

Lemma (11.7). *If* $k \equiv 0 \bmod 4$ *then* $P_k(C)$ *is a* J'-*space. If* $k \equiv 2 \bmod 4$ *then either* $P_k(C)$ *is a* J'-*space or* $\ker \theta = Z_2$.

This time we consider the cofibration

$$P_{k-1}(C) \xrightarrow{i} P_k(C) \xrightarrow{j} P_{k,1}(C)$$

which gives rise to the following diagram.

$$\begin{array}{ccccccc}
J''(P_{k,1}(C)) & \xrightarrow{j^*} & J''(P_k(C)) & \xrightarrow{i^*} & J''(P_{k-1}(C)) & \rightarrow & 0 \\
\downarrow \theta & & \downarrow \theta & & \downarrow \theta & & \\
J'(P_{k,1}(C)) & \xrightarrow{j^*} & J'(P_k(C)) & \xrightarrow{i^*} & J'(P_{k-1}(C)) & &
\end{array}$$

The upper row is exact, by (11.5), and the right-hand θ is an isomorphism, by (11.6). Hence the kernel of the central θ is contained in the image of the upper j^*. Now $P_{k,1}(C)$ is a $(2k - 2)$-sphere, and so

$\tilde{K}(P_{k,1}(C))$ is Z_2 or zero according as $k \equiv 2$ or $0 \bmod 4$. This proves (11.7). When $k \equiv 2 \bmod 4$ it can be shown, as in §6 of [5], that $R_{\#}\beta^{k-1}$ is J'-trivial but not J"-trivial; thus $R_{\#}\beta^{k-1}$ generates $\ker \theta = Z_2$.

This complication does not arise in the quaternionic case, where a straightforward induction with reference to the cofibration

$$P_{k-1}(H) \to P_k(H) \to P_{k,1}(H) = S^{4k-4}$$

leads immediately to

Theorem (11.8). If $k \geq 2$ then $P_k(H)$ is a J'-space.

We are now ready to determine the J-order b_k of $\beta \in \tilde{K}_C(P_k(C))$ and the J-order c_k of $\gamma \in \tilde{K}_C(P_k(H))$. I assert that

$$(11.9) \quad b_k = b'_k, \quad c_k = c'_k,$$

for all values of k, where b'_k and c'_k are the numbers defined in §10. Using the main theorem of J-theory this follows from (11.6), (11.7) in the complex case when $k \not\equiv 2 \bmod 4$, and from (11.8) in the quaternionic case. For the complex case when $k \equiv 2 \bmod 4$, write $k = 2l$, where l is odd. Then $b_k = b'_k$ or $2b'_k$, by (11.7). But $b_k/2$ divides $c_l = c'_l$ and $c'_l = b'_k/2$, as noted at the end of the last section. Therefore $b_k = b'_k$ in this case also and the proof of (11.9) is complete. Hence and from (7.2) we obtain

Theorem (11.10). The complex Stiefel manifold $W_{n,k}$ admits a cross-section if and only if $n \equiv 0 \bmod b_k$. The quaternionic Stiefel manifold $X_{n,k}$ admits a cross-section if and only if $n \equiv 0 \bmod c_k$.

Here b_k and c_k, after (11.9), are given by (10.12) and (10.20). Another application of the main theorem of J-theory is to prove

Theorem (11.11). If the fibration $W_{n,k} \to W_{n,1}$ admits a cross-section then it admits a homotopy-equivariant cross-section.

Consider, to start with, any pointed Z_2-space X. The mapping torus $\hat{X}$ of X contains, as a retract, the mapping torus S^1 of the basepoint. Choose the obvious retraction $r : \hat{X} \to S^1$, which is constant on X, and consider the diagram shown below, where q is the natural projection and i, j are inclusions.

$$\begin{array}{ccccc} & & \tilde{K}_R(S^1) & & \\ & & \downarrow r^* & & \\ \tilde{K}_R(X/S^1) & \xrightarrow{q^*} & \tilde{K}_R(\hat{X}) & \xrightarrow{i^*} & \tilde{K}_R(S^1) \\ & & \downarrow j^* & & \\ & & \tilde{K}_R(X) & & \end{array}$$

Since $i^*r^* = 1$ and $j^*r^* = 0$ it follows at once that $\operatorname{im} j^* = \operatorname{im} j^*q^*$. Now consider the exact sequence

$$\tilde{K}_R(SX) \to \tilde{K}_R(\hat{X}/S^1) \xrightarrow{j^*q^*} \tilde{K}_R(X),$$

where SX is identified with $\hat{X}/(S^1 \vee X)$. If j^* is onto and $\tilde{K}_R(SX) = 0$ then j^*q^* is an isomorphism, and hence

(11.12) $\quad j^*q^* : J(\hat{X}/S^1) \approx J(X),$

by the main theorem of J-theory.

To prove (11.11) we take $X = P_k = P_k(C)$, with Z_2 acting by complex conjugation. The realification $R_{\#}L$ of the complex line bundle over P_k admits Z_2-structure, as before, and the mapping torus of this Z_2-vector bundle provides an extension of $R_{\#}L$ over $\hat{P}_k$. Since $[R_{\#}L]-2$ generates $\tilde{K}_R(P_k)$, by (10.3), it follows that j^* is onto in this case. Moreover $\tilde{K}_R(SP_k) = 0$, by induction on k, and so both our conditions are fulfilled. Any even multiple of $\widehat{R_{\#}L}$, as a Z_2-vector bundle, lies in the image of q^*; hence the J/Z_2-order of such a bundle is equal to the J-order of the corresponding multiple of L. In other words $\hat{b}_k = b_k$, and now (11.10) follows at once from (7.5).

12·The fibre suspension

Let X be a pointed space and let $p : E \to X$ be a fibration with fibre F. By the q-fold fibre suspension ($q = 1, 2, \ldots$) we mean the space $\Sigma^q E$ obtained from

$$(B^q \times E) + (S^{q-1} \times X)$$

by identifying points of $S^{q-1} \times E$ with their images under $1 \times p$. Since $1 \times p$ is onto, points of $\Sigma^q E$ can always be represented by pairs (a, v), where $a \in B^q$, $v \in E$. Under fairly general conditions $\Sigma^q E$ fibres over X with projection given by $(a, v) \mapsto pv$ and fibre the unreduced q-fold suspension of F. The definition can be modified by collapsing to a point the image of $B^q \times e$ in $\Sigma^q E$ where $e \in F$ is basepoint, so that the fibre becomes $S^q F$, the ordinary (reduced) q-fold suspension. In what follows it is convenient to use the modified form of the definition.

The structural maps of the fibre suspension are denoted by

$$B^q \times E \xrightarrow{i} \Sigma^q E \xleftarrow{j} S^{q-1} \times X.$$

Any map $f : (B^r, S^{r-1}) \to (X, x_0)$ $(r \geq 1)$ determines a map

$$f' : B^q \times S^{r-1} \cup S^{q-1} \times B^r \to \Sigma^q E,$$

where f' is constant on $B^q \times S^{r-1}$ and maps $S^{q-1} \times B^r$ according to $j(1 \times f)$. We describe f' as obtained from f by the fibre construction. The transformation $f \mapsto f'$ determines a homomorphism

$$\Gamma : \pi_r(X) \to \pi_{q+r-1}(\Sigma^q E).$$

When $q = 1$, in particular, it is clear from first principles that

$$(12.1) \quad \Gamma = s_{-*} - s_{+*},$$

where $s_\pm : X \to \Sigma E$ is given by $s_\pm(x) = (\pm 1, x)$. I assert that for all values of q we have

$$(12.2) \quad \Gamma = -u_* S^q_* \Delta,$$

as shown below, where Δ is the transgression operator for the original fibration and $u : S^q F \subset \Sigma^q E$.

$$\pi_r(X) \xrightarrow{\Delta} \pi_{r-1}(F) \xrightarrow{S^q_*} \pi_{q+r-1}(S^q F) \xrightarrow{u_*} \pi_{q+r-1}(\Sigma^q E).$$

This implies that if $S^q F$ is a retract of $\Sigma^q E$ with retraction ρ then

$$(12.3) \quad \rho_* \Gamma = -S^q_* \Delta.$$

Here $\rho_* \Gamma$ is given by the transformation $f \mapsto f''$, where f is as before and

$$f'' : B^q \times S^{r-1} \cup S^{q-1} \times B^r \to S^q F$$

is obtained from

$$\rho j(1 \times f) : S^{q-1} \times X \to S^q F$$

by the Hopf construction.

To prove (12.2) (the case $q = 1$ can be found in [72]), choose a map $g : (B^r, S^{r-1}) \to (E, F)$, and let

$$h : B^q \times B^r \to \Sigma^q E$$

be defined by composing $1 \times g$ with the identification map. Note that h is constant on $S^{q-1} \times S^{r-1}$, maps $B^q \times S^{r-1}$ into $S^k F$, and maps $S^{q-1} \times B^r$ into $S^{q-1} \times X$. Thus a homotopy

$$h_t : B^q \times S^{r-1} \cup S^{q-1} \times B^r \to \Sigma^q E$$

is given, for all $x \in S^{q-1}$, $y \in S^{r-1}$, by

$$h_t(ax, y) = h(ax, ty) \quad (0 \leq a \leq 1),$$

$$h_t(x, by) = \text{basepoint} \quad (0 \leq b \leq t)$$

$$= h(x, (1 - b + t)y) \quad (t \le b \le 1).$$

Now $h_1 = u(S^q g_0)$, where $g_0 = g | S^{r-1}$, while $h_0 = f'k$, where k is a map of degree -1 and f' is obtained by the fibre construction, as above, from the map

$$f = pg : (B^r, S^{r-1}) \to (E, F).$$

Since $h_0 \simeq h_1$ this proves (12. 2).

For example, take $p : O_{n,k+1} \to O_{n,k}$ with fibre S_{n-k}. Then $\Sigma^{dk} O_{n,k+1}$ can be formed from

$$(B_k \times O_{n,k+1}) + (S_k \times O_{n,k})$$

by identifying points of $S_k \times O_{n,k+1}$ with their images under $1 \times p$. A retraction

$$\rho : \Sigma^{dk} O_{n,k+1} \to S_n = \Sigma^{dk} S_{n-k}$$

is defined on $B_k \times O_{n,k+1}$ by

$$(12.4) \quad \rho i((x_1, \ldots, x_k), (v_1, \ldots, v_{k+1})) = (x_1 v_1 + \ldots + x_k v_k + (1 - |x|^2)^{\frac{1}{2}} v_{k+1}),$$

where $x = (x_1, \ldots, x_k) \in B_k$, and on $S_k \times O_{n,k}$ by

$$(12.5) \quad \rho j((x_1, \ldots, x_k), (v_1, \ldots, v_k)) = (x_1 v_1 + \ldots + x_k v_k),$$

where $x = (x_1, \ldots, x_k) \in S_k$. To calculate

$$-S_*^{dk} \Delta : \pi_r(O_{n,k}) \to \pi_{r+dk-1}(S_n),$$

therefore, we choose a representative $f : S^r \to O_{n,k}$ and apply the Hopf construction to

$$\rho j(1 \times f) : S_k \times S^r \to S_n,$$

where ρj is given by (12. 5).

In particular, consider the real case with $n \equiv 0 \bmod a_k$. Choose an orthogonal C_{k-1}-module structure on R^n. This determines both a Clifford cross-section $S^{n-1} \to V_{n,k}$ and a map $S^{k-1} \to O_n$, in the obvious way. If $\sigma \in \pi_{n-1}(V_{n,k})$ and $\sigma' \in \pi_{k-1}(O_n)$ are the classes of these maps our formula shows that

$$(12.6) \quad S_*^k \Delta\sigma = \pm J\sigma'.$$

This generalizes the results obtained at the end of §5 regarding the cross-sections given by the inclusions

$$W_{m,1} \subset V_{2m,2}, \quad X_{m,1} \subset V_{4m,4}.$$

Note that if $n = a_k$, so that R^n is irreducible, then $J\pi_{k-1}(O_n)$ defines a cyclic summand of known order in the stable group of the (k - 1)-stem (see [4] for details).

Let us briefly digress and prove

Theorem (12.7). <u>The Stiefel manifold</u> $O_{n,k}$ <u>is stably parallelizable.</u>

This result is the first stage in Sutherland's proof [139] that $O_{n,k}$ is parallelizable when $k \geq 2$. To prove (12.7) recall that if V is a smooth vector bundle over a smooth manifold M then

$$(12.8) \quad 1 \oplus T(S(V)) \approx p^*(V \oplus T(M)),$$

where T denotes the tangent bundle and $p : S(V) \to M$ the projection. Thus if V is stably trivial and M is stably parallelizable then $S(V)$ is stably parallelizable. Now the retraction $\Sigma^{dk} O_{n,k+1} \to S_n$ defined above determines a trivialization of $\Sigma^{dk} O_{n,k+1}$ as a sphere-bundle over $O_{n,k}$. Hence (12.7) follows by induction.

In view of (12.7) the duality theorem of Milnor and Spanier [113] (see also Atiyah [7]) shows that $O_{n,k}$ is self-dual, in the sense of S-theory. Because our procedure uses the specific trivialization determined by the retraction in (12.4) we obtain, in this way, a specific class of duality map, so that the dual of a class of self-maps of $O_{n,k}$ is well-defined. By using a homotopy-equivariant version of the above argument, and ignoring changes of sign, it is easily shown that the classes of λ and μ are both self-dual in the real case, and the class of complex conjugation is self-dual in the complex case.

Since $O_{n,k}$ is self-dual and $Q_{n,k}$ is an S-retract of $O_{n,k}$, as we have seen, it follows at once that the dual of $Q_{n,k}$ (suitably suspended) is also an S-retract of $O_{n,k}$. It appears, therefore, that the S-type of $O_{n,k}$ splits into the wedge-sum of four spaces: a stunted quasi-projective space at the bottom, a middle section (about which little is known), a stunted projective space next to the top, and finally a sphere in the top dimension.

13·Canonical automorphisms

In this section we study the canonical classes λ, μ and $\xi = \lambda\mu$ of self-maps of $V_{n,k}$ defined in §1 and particularly the induced automorphisms of $\pi_r(V_{n,k})$. After showing that

(13.1) $\xi_* = 1 - \Delta S_* p_*$,

where

$$\pi_r(V_{n,k}) \xrightarrow{p_*} \pi_r(S^{n-1}) \xrightarrow{S_*} \pi_{r+1}(S^n) \xrightarrow{\Delta} \pi_r(V_{n,k}),$$

we shall use the results of §12 to show that

(13.2) $\lambda_* = 1 - u_* S_* \Delta$,

where

$$\pi_r(V_{n,k}) \xrightarrow{\Delta} \pi_{r-1}(S^{n-k-1}) \xrightarrow{S_*} \pi_r(S^{n-k}) \xrightarrow{u_*} \pi_r(V_{n,k}).$$

Since $S_*\Delta\mu_* = S_*\mu_*\Delta = -S_*\Delta$ we deduce from (13.1) and (13.2) that

(13.3) $\mu_* = 1 - u_* S_* \Delta - \Delta S_* p_*$.

These formulae can be used to obtain information about the transgression operator, as well as the induced automorphisms. For example, recall that

(13.4) $\lambda_*^k = \mu_*^n$,

from (1.1). Using this we deduce from (13.2) that $u_* S_* \Delta = 0$ when n is even and k is odd, from (13.1) that $\Delta S_* p_* = 0$ when n and k are both odd, and from (13.3) that $u_* S_* \Delta = -\Delta S_* p_*$ when n is odd and k is even.

In the stable range, when $V_{n,k}$ can be replaced by $P_{n,k}$, we can deduce (13.1) from (13.2) by using the duality formulae of §7. To prove the identity in general, however, we consider the following situation. Let H be a subgroup of a topological group G and G/H the factor space of left cosets. The space $\Omega(G, H)$ of paths in G from the neutral element e to H inherits the structure of a topological group from G. Hence the space of loops on $\Omega(G, H)$ inherits a group-structure, which of course is homotopically equivalent to the H-structure given by loop composition. Hence the group structure in $\pi_r(G, H)$ $(r \geq 2)$ determined by the topological group structure of the pair (G, H) coincides with the ordinary group structure given by track addition.

Suppose now that we have a self-inverse automorphism ξ of G which acts as the identity on H, and let ξ also denote the induced involution of G/H. Consider the composition

$$(G, H) \xrightarrow{\rho} (G/H, e) \xrightarrow{\zeta} (G, H),$$

where ρ is the natural projection and

$$\zeta\rho(g) = g. \xi(g^{-1}) \qquad (g \in G).$$

Then $\zeta_*\rho_* = 1 - \xi_*$, by the above remarks, where ξ_* denotes the automorphism of $\pi_r(G, H)$ induced by ξ. Now suppose that

$$\rho_* : \pi_r(G, H) \to \pi_r(G/H)$$

is an isomorphism, as is the case (see [133]) when (G, H) is a Lie pair. Since

$$\rho_*\zeta_*\rho_* = \rho_* - \rho_*\xi_* = \rho_* - \xi_*\rho_*,$$

we obtain the relation

$$(13.5) \quad \rho_*\zeta_* = 1 - \xi_*.$$

To prove (13.1) take $(G, H) = (O_n, O_{n-k})$, with ξ defined by conjugation by the matrix

$$1 \oplus 1 \oplus \dots \oplus 1 \oplus -1 \qquad (n \text{ summands}).$$

A straightforward calculation, with reference to (23.3) of [133], shows that the self-map $\rho\zeta$ of $V_{n,k}$ coincides with the composition

$$V_{n,k} \xrightarrow{p} S^{n-1} \xrightarrow{d} V_{n,k},$$

where d is the projection of the classifying map $S^{n-1} \to O_n$. Since d represents $\Delta\iota_n$, where

$$\Delta : \pi_{r+1}(S^n) \to \pi_r(V_{n,k}),$$

it follows that $d_* = \Delta S_*$ and hence that

$$\Delta S_* p_* = d_* p_* = \rho_* \zeta_* = 1 - \xi_*,$$

by (13.5). This proves (13.1).

To establish our second result we generalize in a different way and use the main theorem of §12. Let X be a fibre space over a fibre space Y over a space Z, as shown in the following diagram.

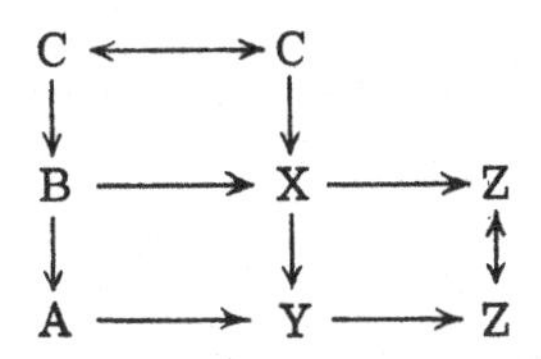

Here A is the fibre of Y over Z, B is the fibre of X over Z, and C is the fibre of X over Y. Thus each of the horizontal and vertical rows of three forms a triple (fibre, total space, base).

Let $f_\pm : Y \to Y$ be a pair of fibre-preserving maps, over Z, and let $h_t : X \to Y$ $(-1 \le t \le 1)$ be a fibre-preserving homotopy, over Z, such that $h_{\pm 1} = f_\pm p$, where $p : X \to Y$ is the fibration. Now ΣX, the fibre-suspension of X, is a fibre space over Y with fibre SC and canonical cross-sections $s_\pm : Y \to \Sigma X$. Consider the map $g : \Sigma X \to Y$ which is given by

$$gi(t, x) = h_t x \quad (x \in X,\ t \in I),$$
$$gj(\pm 1, y) = f_\pm y \quad (y \in Y).$$

By (12.1) and (12.2) we have the relations

$$s_{+*} - s_{-*} = u_* S_* \Delta,$$

where

$$\pi_r(Y) \xrightarrow{\Delta} \pi_{r-1}(C) \xrightarrow{S_*} \pi_r(SC) \xrightarrow{u_*} \pi_r(\Sigma X).$$

Since $gs_\pm = f_\pm$ it follows at once that

$$f_{+*} - f_{-*} = g_* u_* S_* \Delta.$$

But $gu = vg_0$, where $v : A \subset Y$ and where $g_0 : SC \to A$ is determined by g. Therefore

$$(13.6) \quad f_{+*} - f_{-*} = v_* g_{0*} S_* \Delta,$$

where

$$\pi_r(Y) \xrightarrow{\Delta} \pi_{r-1}(C) \xrightarrow{S_*} \pi_r(SC) \xrightarrow{g_{0*}} \pi_r(A) \xrightarrow{v_*} \pi_r(Y).$$

To establish (13.2) we apply this formula in relation to the diagram shown below.

$$\begin{array}{ccccc} S^{n-k-1} & \longrightarrow & S^{n-k-1} & & \\ \downarrow & & \downarrow & & \\ V_{n-k+1,2} & \longrightarrow & V_{n,k+1} & \longrightarrow & V_{n,k-1} \\ \downarrow & & \downarrow & & \downarrow \\ S^{n-k} & \longrightarrow & V_{n,k} & \longrightarrow & V_{n,k-1} \end{array}$$

Take f_+ to be the identity on $V_{n,k}$ and f_- the self-map which changes the sign of the first vector in each k-frame. Then a homotopy $h_t : V_{n,k+1} \to V_{n,k}$ satisfying our requirements, is given by

$$h_t(v_1, v_2, v_3, \ldots, v_{k+1}) = (v_1 \sin\tfrac{\pi}{2}t + v_2 \cos\tfrac{\pi}{2}t, \; v_3, \ldots, v_{k+1}).$$

In this case the map $g_0 : S(S^{n-k-1}) \to S^{n-k}$ has degree 1, and so (13.6) yields

$$1 - \lambda_* = u_* S_* \Delta,$$

since f_- is of class λ. This proves (13.2).

For example, consider the homotopy class $\beta_m \in \pi_{2m-1}(V_{2m,2})$ $(m = 1, 2, \ldots)$ of the inclusion $W_{m,1} \subset V_{2m,2}$. From (5.8) we at once obtain that $\Delta\beta_m = m\eta$, and so

$$(13.7) \quad \lambda_*\beta_m = \beta_m - mu_*\eta,$$

by (13.2). If $m \geq 2$ then $\beta_m = \beta_1 * \beta_{m-1}$ and so

$$\xi_*(\beta_m) = \xi_*(\beta_1 * \beta_{m-1}) = (\xi_*\beta_1) * (\lambda_*\beta_{m-1}),$$

by (2.4). Since $\xi_*\beta_1 = -\beta_1$, from degree considerations, and $\beta_1 * u_*\eta_{2m-4} = u_*\eta_{2m-2}$, as in the generalized Freudenthal theorem, this shows that

$$(13.8) \quad \xi_*\beta_m = -\beta_m + (m-1)u_*\eta,$$

and hence

$$(13.9) \quad \mu_*\beta_m = -\beta_m + u_*\eta.$$

Also $\Delta\iota_{2m} = 2\beta_m - (m-1)u_*\eta$, by (13.1), where

$$\Delta : \pi_{r+1}(S^{2m}) \to \pi_r(V_{2m,2}).$$

These formulae can be used to calculate the action of λ_* and μ_* on $\pi_r(V_{2m,2})$ for all values of r.

Similar calculations can be made in respect of $\gamma_m \in \pi_{4m-1}(V_{4m,4})$, the homotopy class of $X_{m,1} \subset V_{4m,4}$. By (5.9) and (13.2) we have that

$$(13.10) \quad \lambda_*\gamma_m = \gamma_m - mu_*\nu,$$

for $m \geq 3$. Recall (see §22 of [133]) that $\pi_3(O_4)$ is freely generated by the pair (α, β), where $\beta = \gamma_1$ and $\alpha = \beta + \xi_*\beta$. If $m \geq 2$ then $\gamma_m = \beta * \gamma_{m-1}$ and so

$$(13.11) \quad \xi_*\gamma_m = \delta_m - \gamma_m + (m-1)u_*\nu,$$

where $\delta_m = \alpha * \gamma_{m-1}$, and therefore

$$(13.12) \quad \mu_*\gamma_m = \delta_m - \gamma_m - u_*\nu.$$

Comparing (13. 11) with (13. 1) we see that

$$(13.13) \quad \Delta \iota_{4m} = 2\gamma_m - \delta_m - (m - 1)u_* \nu,$$

where $\Delta : \pi_{r+1}(S^{4m}) \to \pi_r(V_{4m, 4})$. I do not know of any other way to obtain this result. These formulae can be used to calculate the action of λ_* and μ_* on $\pi_r(V_{4m, 4})$, for all values of r.

14 · Iterated suspension

The purpose of this section is to show how the homotopy groups of Stiefel manifolds and stunted projective spaces play a special role in the study of the homotopy groups of spheres.

Let H_n $(n = 1, 2, \ldots)$ denote the space of (free) self-maps of S^{n-1} which are homotopy equivalences, with the obvious embedding $O_n \subset H_n$. Let $p : H_n \to S^{n-1}$ be defined by evaluation at the basepoint. Then p is a fibration with fibre the subspace F_{n-1} of pointed maps. We embed H_n in F_n, using the suspension functor, and consider the homotopy exact sequence of the triple (F_n, H_n, F_{n-1}), as follows:

$$\ldots \to \pi_r(H_n, F_{n-1}) \xrightarrow{i_*} \pi_r(F_n, F_{n-1}) \xrightarrow{j_*} \pi_r(F_n, H_n) \to \ldots .$$

Since p is a fibration we have that

$$p_* : \pi_r(H_n, F_{n-1}) \approx \pi_r(S^{n-1}),$$

for all values of r. From first principles we have a generalized Hurewicz isomorphism

$$\theta : \pi_r(F_n, F_{n-1}) \approx \pi_{r+n}(S^n; B^n_+, B^n_-),$$

where B^n_+, B^n_- denote the closed hemispheres into which S^{n-1} divides S^n. Choose a generator $\kappa_n \in \pi_n(B^n_-, S^{n-1})$ and consider the boundary isomorphism

$$\delta : \pi_{r+1}(B^n_+, S^{n-1}) \approx \pi_r(S^{n-1}).$$

Then a homomorphism ϕ is defined, as shown below, so that if $\alpha \in \pi_r(S^{n-1})$ then $\phi(\alpha) = [\delta^{-1}\alpha, \kappa_n]$, the triad Whitehead product.

$$\begin{array}{ccc} \pi_r(H_n, F_{n-1}) & \xrightarrow{i_*} & \pi_r(F_n, F_{n-1}) \\ p_* \downarrow & & \downarrow \theta \\ \pi_r(S^{n-1}) & \xrightarrow[\phi]{} & \pi_{r+n}(S^n; B^n_+, B^n_-) \end{array}$$

An unpublished result of W. E. Sutherland's asserts that

(14. 1) $\phi p_* = \pm\theta i_*$.

Assuming this we deduce

Theorem (14. 2). _The pair_ (F_n, H_n) _is_ $(2n - 3)$-_connected._

This follows at once from (14. 1) and the main theorem of [22] and [143] on the triad Whitehead product. The proof originally given in [63] loses a dimension. However Haefliger [51] has given an entirely different proof, using the methods of differential topology, and goes much further by establishing

Theorem (14. 3). _If_ $r \leq 3n - 6$ _then_

$$\pi_r(F_n, H_n) \approx \pi_{r-n+1}(O, O_{n-1}).$$

Here O, of course, denotes the stable orthogonal group.

The inclusions $(O_{n+1}, O_n) \subset (H_{n+1}, F_n) \subset (F_{n+1}, F_n)$ induce homomorphisms

$$\pi_r(O_{n+1}, O_n) \xrightarrow{u_*} \pi_r(H_{n+1}, F_n) \xrightarrow{v_*} \pi_r(F_{n+1}, F_n).$$

The fibrations $p : H_{n+1} \to S^n$ and $p' = p|O_{n+1}$ induce isomorphisms

$$\pi_r(O_{n+1}, O_n) \xrightarrow{p'_*} \pi_r(S^n) \xleftarrow{p_*} \pi_r(H_{n+1}, F_n).$$

Therefore u_* is an isomorphism, since $p'_* = p_*u_*$, while v_* is $(2m - 2)$-connected, from (14. 2). Hence if $r \leq 2n - 2$ then

$$v_*u_* : \pi_r(O_{n+1}, O_n) \approx \pi_r(F_{n+1}, F_n).$$

Hence by induction on k, using the five lemma, we obtain

Theorem (14.4). If $r \leq 2n - 2$ then

$$w_* : \pi_r(O_{n+k}, O_n) \approx \pi_r(F_{n+k}, F_n),$$

where $w : (O_{n+k}, O_n) \subset (F_{n+k}, F_n)$.

Recall (see [152], [160]) that the Hurewicz isomorphism θ satisfies the relation $\theta i_* = \pm S_*^k \theta$, as shown in the following diagram.

$$\begin{array}{ccc} \pi_r(F_n) & \xrightarrow{i_*} & \pi_r(F_{n+k}) \\ \downarrow \theta & & \downarrow \theta \\ \pi_{r+n}(S^n) & \xrightarrow[S_*^k]{} & \pi_{r+n+k}(S^{n+k}) \end{array}$$

Write $P_k = J\Delta$, where

$$\pi_r(V_{n+k,k}) \xrightarrow{\Delta} \pi_{r-1}(O_n) \xrightarrow{J} \pi_{r+n-1}(S^n),$$

and if $r \leq 2n - 2$, so that (14.4) applies, write $H_k = w_*^{-1} j_* \theta^{-1}$, where

$$\pi_{r+n+k}(S^{n+k}) \xleftarrow{\theta} \pi_r(F_{n+k}) \xrightarrow{j_*} \pi_r(F_{n+k}, F_n) \xleftarrow{w_*} \pi_r(O_{n+k}, O_n).$$

Then (14.4) implies that the sequence

$$(14.5) \quad \pi_{2n-2}(S^n) \to \dots$$

$$\dots \to \pi_{r+n}(S^n) \xrightarrow{S_*^k} \pi_{r+n+k}(S^{n+k}) \xrightarrow{H_k} \pi_r(V_{n+k,k}) \xrightarrow{P_k} \pi_{r+n-1}(S^n) \to \dots$$

is exact. This can be regarded as a generalization of the EHP sequence of G. W. Whitehead [155]; an application will be given in §21 below.

Another treatment of the iterated suspension, with certain advantages, is due to Toda [148]. We describe this briefly. Consider the action $O_n \times S^{n-1} \to S^{n-1}$ of O_n on S^{n-1}. Applying the Hopf construction we obtain a map $h : S^n O_n \to S^n$ with the property that $h_* S_*^n = J$, where

$$\pi_i(O_n) \xrightarrow{S_*^n} \pi_{i+n}(S^n O_n) \xrightarrow{h_*} \pi_{i+n}(S^n).$$

The maps h can be chosen so as to be compatible for various values of n. Hence the adjoint of h determines a map

$$f : (S^n O_{n+k}, S^n O_n) \to (\Omega^k(S^{n+k}), S^n),$$

for $k = 1, 2, \ldots$. When $k = 1$ the composition

$$\pi_n(O_{n+1}, O_n) \xrightarrow{S^n_*} \pi_{2n}(S^n O_{n+1}, S^n O_n) \xrightarrow{f_*} \pi_{2n}(\Omega(S^{n+1}), S^n)$$

is easily shown to be an isomorphism, from which it follows that

$$g_* : \pi_{2n}(S^n P_{n+1}, S^n P_n) \approx \pi_{2n}(\Omega(S^{n+1}), S^n),$$

where $g = f | S^n(P_{n+1}, P_n)$.

Given any pair (X, A) consider the space $X \cup CA$ obtained by attaching the cone CA on A to X. Each point $x \in X$ determines a loop in SX, given by $t \mapsto (x, t)$. Each point $(a, s) \in CA$ determines a path in SA, given by $t \mapsto (a, st + 1 - s)$. These transformations agree on A and so determine a map i of $X \cup CA$ into $\Omega(SX, SA)$, the space of paths in SX which start in SA and end at the basepoint. When (X, A) is a CW-pair the natural projection $X \cup CA \to X/A$ is a homotopy equivalence, and by composing i with a homotopy inverse of this we obtain a class of maps $j : X/A \to \Omega(SX, SA)$.

Consider, in particular, such a map

$$j : S^{n-1}P_{n+k, k} \to \Omega(S^n P_{n+k}, S^n P_n).$$

Composing this with the restriction of

$$\Omega f : \Omega(S^n O_{n+k}, S^n O_n) \to \Omega(\Omega^k(S^{n+k}), S^n)$$

we obtain a map

$$e : S^{n-1}P_{n+k, k} \to \Omega(\Omega^k(S^{n+k}), S^n).$$

Toda refers to the induced homomorphism

$$e_* : \pi_r(S^{n-1}P_{n+k, k}) \to \pi_{r+1}(\Omega^k(S^{n+k}), S^n)$$

as _the generalized_ J-_homomorphism,_ and proves

Theorem (14.6). _If_ $r < 2n - 2$ _then_ e_*, _as above, is an iso-_

morphism. If $r < 4n - 3$ then e_* is an isomorphism of 2-primary components.

This is proved by induction on k, using what is essentially a five lemma argument based on the case $k = 1$, which we have already dealt with. The memoir of Toda [148] is the basic reference but the recent note by Nomura [115] is also relevant. This line of thought leads on to the important result of Barratt and Mahowald [17], which states that if $n \leq 13$ and $r \leq 2n - 3$ then the homotopy exact sequence

$$\pi_{r+1}(O, O_n) \to \pi_r(O_n) \to \pi_r(O)$$

is short exact and splits. Further information about this, with calculations in the range $r - n \leq 29$, can be found in the memoir of Mahowald [100] (see also [58]).

15 · Samelson products

Following Leise [95] our treatment of Samelson products is based on the following identities, valid for any elements a, b, c of a group G with neutral element e:

$$(15.1)\begin{cases}[a, b][b, a] = e,\\ [a, bc] = [a, b][a, c][[c, a], b]\\ [a^c, [b, c]][c^b, [a, b]][b^a, [c, a]] = e.\end{cases}$$

Here $[a, b] = aba^{-1}b^{-1}$ and $a^b = bab^{-1}$.

For any topological group G the (ordinary) Samelson product $\langle , \rangle$ is a pairing

$$\pi_p(G) \times \pi_q(G) \to \pi_{p+q}(G) \qquad (p, q \geq 1)$$

defined as follows. Let $\alpha \in \pi_p(G)$, $\beta \in \pi_q(G)$ be represented by maps

$$(I^p, \dot{I}^p) \xrightarrow{f} (G, e) \xleftarrow{g} (I^q, \dot{I}^q).$$

Then $\langle \alpha, \beta \rangle \in \pi_{p+q}(G)$ is defined to be the element represented by

$$h : (I^p \times I^q, \dot{I}^p \times I^q \cup I^p \times \dot{I}^q) \to (G, e),$$

where $h(x, y) = [fx, gy]$ $(x \in I^p, y \in I^q)$. The commutation law

$$(15.2) \quad \langle \alpha, \beta \rangle = (-1)^{pq-1}\langle \beta, \alpha \rangle$$

follows at once from the first identity in (15.1), while bilinearity can easily be deduced from the second. Notice that if G' is a topological group and $\phi : G \to G'$ a homomorphism then

$$(15.3) \quad \phi_*\langle \alpha, \beta \rangle = \langle \phi_*\alpha, \phi_*\beta \rangle,$$

where $\phi_* : \pi_*G \to \pi_*G'$ is the induced homomorphism. Moreover if

$\xi \in \pi_i(S^p)$, $\eta \in \pi_j(S^q)$ then

$$(15.4) \quad \langle \alpha \circ \xi, \beta \circ \eta \rangle = \langle \alpha, \beta \rangle \circ (\xi \wedge \eta),$$

where $\xi \wedge \eta \in \pi_{i+j}(S^{p+q})$ denotes the smash product in the homotopy groups of spheres. This can be seen at once by recasting the definition in terms of maps of spheres.

The Samelson product satisfies the following form of Jacobi identity:

$$(15.5) \quad (-1)^{pr}\langle \alpha, \langle \beta, \gamma \rangle\rangle + (-1)^{rq}\langle \gamma, \langle \alpha, \beta \rangle\rangle + (-1)^{pq}\langle \beta, \langle \gamma, \alpha \rangle\rangle = 0,$$

where $\alpha \in \pi_p(G)$, $\beta \in \pi_q(G)$, $\gamma \in \pi_r(G)$ $(p, q, r \geq 1)$. There are various proofs in the literature of which the following, due to Leise, is particularly simple. Choose representatives f, g, h of α, β, γ so that $f : (I^p, \dot{I}^p) \to (G, e)$, etc. Consider the homotopies

$$k_t, l_t, m_t : (I^{p+q+r}, \dot{I}^{p+q+r}) \to (G, e) \qquad (t \in I)$$

defined for $x \in I^p$, $y \in I^q$, $z \in I^r$ by

$$(15.6) \quad \begin{cases} k_t(x, y, z) = [h(tz)f(x)h^{-1}(tz), [gy, hz]], \\ l_t(x, y, z) = [g(ty)h(z)g^{-1}(ty), [fx, gy]], \\ m_t(x, y, z) = [f(tx)g(y)f^{-1}(tx), [hz, fx]]. \end{cases}$$

When $t = 0$ these three maps represent

$$\langle \alpha, \langle \beta, \gamma \rangle\rangle, \quad (-1)^{(p+q)r}\langle \gamma, \langle \alpha, \beta \rangle\rangle, \quad (-1)^{p(r+q)}\langle \beta, \langle \gamma, \alpha \rangle\rangle$$

respectively. When $t = 1$, however, the product $k_1 l_1 m_1$ is constant, by the third identity in (15.1). This proves (15.5).

Another way to look at the Jacobi identity is to consider the operator

$$\gamma_\# : \pi_t(G) \to \pi_{r+t}(G) \qquad (t = 1, 2, \ldots)$$

given by the Samelson product with γ. Then (15.5) is equivalent to the derivation law

(15.7) $\gamma_{\#}\langle\alpha, \beta\rangle = \langle\gamma_{\#}\alpha, \beta\rangle + (-1)^{qr}\langle\alpha, \gamma_{\#}\beta\rangle$.

In particular, take $r = 1$ and suppose that $\pi_2(G) = 0$, as is the case when G is a Lie group. Then $2\gamma_{\#}^2 = 0$; moreover there is some evidence to support the

Conjecture (15.8). _For some value of_ s, _depending on_ γ _but not on_ r, _the operator_

$$\gamma_{\#}^s : \pi_r(G) \to \pi_{r+s}(G) \qquad (r = 1, 2, \dots),$$

defined by iteration, is trivial.

For example take $G = R_t$, the rotation group. Let $D : \pi_r(R_t) \to \pi_{r+1}(R_t)$ be defined by taking the Samelson product with the generator $\theta \in \pi_1(R_t)$. Take $t > 2$, since $t = 2$ is trivial. We prove

Proposition (15.9). _If_ $t \equiv 2 \bmod 4$ _then_ $D^2 = 0$.

The projective group PR_t is defined to be the quotient of R_t by the centre $\{e, -e\}$. The natural homomorphism $\rho : R_t \to PR_t$ induces a homomorphism

$$\rho_* : \pi_r(R_t) \to \pi_r(PR_t)$$

which respects Samelson products and is an isomorphism for $r \geq 2$. Since $t \equiv 2 \bmod 4$ we have that $\pi_1(PR_t) = Z_4$ with generator ϕ, say, such that $2\phi = \rho_*\theta$. Hence if $\alpha \in \pi_r(R_t)$, where $r \geq 2$, we have that $\rho_*D\alpha = 2\langle\rho_*\alpha, \phi\rangle$ and therefore

$$\rho_*D^2\alpha = 4\langle\langle\rho_*\alpha, \phi\rangle, \phi\rangle = 2\langle\rho_*\alpha, \langle\phi, \phi\rangle\rangle,$$

by the Jacobi identity in $\pi_*(PR_t)$. But $\langle\phi, \phi\rangle = 0$, since $\pi_2(PR_t) = 0$, and so $D^2\alpha = 0$, since ρ_* is an isomorphism. This proves (15.9). Further information about the operator D will be obtained in the next section, after relative Samelson products have been discussed.

Let H be a subgroup of the topological group G. The relative Samelson product $\langle , \rangle$ is a pairing

$$\pi_p(H) \times \pi_q(G, H) \to \pi_{p+q}(G, H)$$

defined as follows, for $p \geq 1$, $q \geq 2$. Let $\alpha \in \pi_p(H)$, $\beta \in \pi_q(G, H)$ be represented by maps

$$f : (I^p, \dot{I}^p) \to (H, e), \quad g : (I^q, \dot{I}^q) \to (G, H) .$$

Then $\langle \alpha, \beta \rangle \in \pi_{p+q}(G, H)$ is represented by the map

$$h : (I^{p+q}, \dot{I}^{p+q}) \to (G, H),$$

where $h(x, y) = [fx, gy]$ $(x \in I^p, y \in I^q)$. For formal reasons it is also desirable to introduce the pairing $\langle , \rangle$:

$$\pi_p(G, H) \times \pi_q(H) \to \pi_{p+q}(G, H) \quad (p \geq 2, q \geq 1),$$

similarly defined. The commutative law

$$(15.10) \quad \langle \alpha, \beta \rangle = (-1)^{pq-1} \langle \beta, \alpha \rangle$$

is easily verified. The basic properties of the relative Samelson product, such as bilinearity, are established just as in the ordinary case. For the composition law, let $\eta' \in \pi_j(B^q, S^{q-1})$ denote the semisuspension of an element $\eta \in \pi_{j-1}(S^{q-1})$. Then if $\xi \in \pi_i(S^p)$ we have

$$(15.11) \quad \langle \alpha \circ \xi, \beta \circ \eta' \rangle = \langle \alpha, \beta \rangle \circ (\xi \wedge \eta)',$$

for $\alpha \in \pi_p(H)$, $\beta \in \pi_q(G, H)$. Leise's proof of the Jacobi identity applies without alteration to the relative case, and shows that

$$(15.12) \quad (-1)^{pr} \langle \alpha, \langle \beta, \gamma \rangle \rangle + (-1)^{rq} \langle \gamma, \langle \alpha, \beta \rangle \rangle + (-1)^{pq} \langle \beta, \langle \gamma, \alpha \rangle \rangle = 0,$$

where $\alpha \in \pi_p(G, H)$ $(p \geq 2)$, $\beta \in \pi_q(H)$ $(q \geq 1)$, and $\gamma \in \pi_r(H)$ $(r \geq 1)$.

The main relations between the ordinary and relative Samelson product are indicated in the following diagrams.

$$(15.13) \quad \begin{array}{ccc} \pi_p(H) \times \pi_q(G, H) & \xrightarrow{\langle , \rangle} & \pi_{p+q}(G, H) \\ {\scriptstyle 1 \times \delta} \downarrow & & \downarrow {\scriptstyle \delta} \\ \pi_p(H) \times \pi_{q-1}(H) & \xrightarrow{\langle , \rangle} & \pi_{p+q-1}(H) \end{array}$$

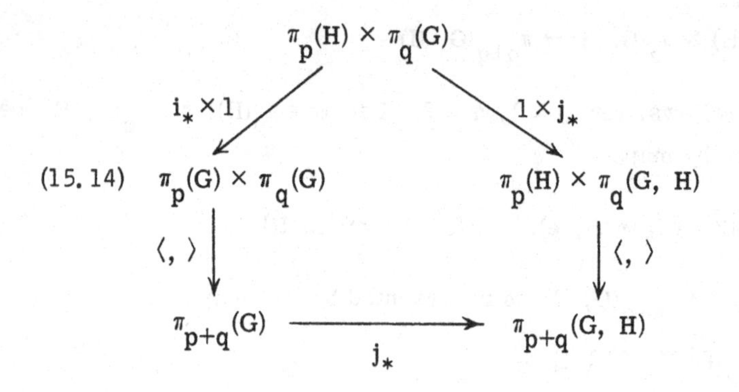

The homomorphisms i_*, j_*, δ, of course, are from the homotopy exact sequence of the pair (G, H), and the diagrams are commutative, apart from sign changes. We see from this that an element $\gamma \in \pi_r(H)$ determines a homomorphism of the homotopy exact sequence into itself, raising dimension by r. On $\pi_*(H)$ we take the ordinary Samelson product with γ itself, on $\pi_*(G)$ the ordinary Samelson product with the image of γ in $\pi_r(G)$, and on $\pi_*(G, H)$ the relative Samelson product with γ itself.

Another approach to Samelson products is to consider the classifying space B_G of G. This is the base space of a principal G-bundle with contractible total space and so the transgression

$$\Delta : \pi_{r+1}(B_G) \approx \pi_r(G) \qquad (r = 1, 2, \ldots)$$

is an isomorphism. Samelson [122] has shown that

$$(15.15) \quad \Delta[\xi, \eta] = \pm\langle \Delta\xi, \Delta\eta \rangle,$$

where $\xi, \eta \in \pi_*(B_G)$. Furthermore Δ can be relativized so as to constitute an isomorphism, of degree -1, from the homotopy exact sequence of the pair (B_G, B_H) to the homotopy exact sequence of the pair (G, H); this isomorphism transforms the Whitehead product structure of the former into the Samelson product structure of the latter.

Finally, consider the homomorphism $\phi_* : \pi_*(G, H) \to \pi_*(G/H)$, induced by the natural projection ϕ onto the factor space. Suppose that ϕ_* is an isomorphism, as is the case when (G, H) is a Lie pair. Then the relative Samelson product can be regarded as a pairing

$$\pi_p(H) \times \pi_q(G/H) \to \pi_{p+q}(G/H).$$

This is the point of view we shall adopt in the next section.

16 · The Hopf construction

Following [80] our approach to the Hopf construction is based on the work of McCarty [96], as well as earlier authors. Let A and Y be pointed spaces. We say that a map $\mu : A \times Y \to Y$ satisfies the McCarty condition if

$$(16.1) \qquad \begin{cases} \mu(a, e) = e & (a \in A), \\ \mu(e, y) = y & (y \in Y). \end{cases}$$

Under this condition a pairing of $\pi_p(A)$ with $\pi_q(Y)$ to $\pi_{p+q}(Y)$, where $p, q \geq 1$, is defined as follows. Let $\alpha \in \pi_p(A)$, $\beta \in \pi_q(Y)$ be represented by $f : S^p \to A$, $g : S^q \to Y$, and let $\rho : S^p \times S^q \to S^q$ denote projection on the second factor. By (16.1) the maps $\mu(f \times g)$ and $g\rho$ coincide on the subspace

$$S^p \times e \cup e \times S^q \subset S^p \times S^q,$$

and so the separation element

$$d(\mu(f \times g), g\rho) \in \pi_{p+q}(Y)$$

is defined. This element, which depends only on α, β and μ, will be denoted by $J_\mu(\alpha, \beta)$.

For example, let G be a topological group and let

$$\mu : G \times G \to G$$

be defined by $\mu(x, y) = xyx^{-1}$. Then $J_\mu(\alpha, \beta) = \langle \alpha, \beta \rangle$, the Samelson product.

For another example, let (G, H) be a Lie pair and let

$$\mu : H \times (G/H) \to G/H$$

be induced by the translation action. Then $J_\mu(\alpha, \beta) = \langle \alpha, \beta \rangle$, the relative Samelson product. In particular, suppose that $G/H = S^q$ and $\beta = \iota_q$. Writing $J_\mu(\alpha, \iota_q) = J\alpha$, in the usual way, we obtain the relation

$$(16.2) \quad \langle \alpha, \iota_q \rangle = J\alpha.$$

Returning to the general case, suppose that $\phi : G \to G/H$ is a fibration with fibre H and transgression $\Delta : \pi_{r+1}(G/H) \to \pi_r(H)$. McCarty [64] has proved that

$$(16.3) \quad [\alpha, \beta] = \pm\langle \Delta\alpha, \beta \rangle$$

where $\alpha, \beta \in \pi_*(G/H)$ and square brackets denote the Whitehead product. To see this, consider the relative loop space $\Omega(G, H)$ of paths in G from e to H. This space inherits from G the structure of a topological group so that the evaluation map $d : \Omega(G, H) \to H$ is a homomorphism. Now

$$\Omega\phi : \Omega(G, H) \to \Omega(G/H)$$

preserves H-structures, up to homotopy, and so the Whitehead product in G/H can be identified (up to sign) with the Samelson product for $\Omega(G, H)$, by Samelson's theorem (15.15). Hence (16.3) follows from consideration of the diagram shown below, where the horizontals are all determined by the group operation of G.

$$\begin{array}{ccc}
\Omega(G, H) \times \Omega(G, H) & \longrightarrow & \Omega(G, H) \\
\downarrow d \times 1 & & \downarrow 1 \\
H \times \Omega(G, H) & \longrightarrow & \Omega(G, H) \\
\downarrow 1 \times \Omega\phi & & \downarrow \Omega\phi \\
H \times \Omega(G/H) & \longrightarrow & \Omega(G/H)
\end{array}$$

Notice, incidentally, that $\langle \Delta\alpha, \beta \rangle = \pm\langle \alpha, \Delta\beta \rangle$.

Another approach to the relative Samelson product is to consider H-bundles over a given base with fibre $Y = G/H$. Since the action of H on Y is pointed each of these bundles possesses a canonical cross-section. When the base is S^n $(n = 2, 3, \ldots)$ the classes of H-bundles correspond to elements of $\pi_{n-1}(H)$. Consider the bundle E with fibre Y over S^n which corresponds to a given element $\alpha \in \pi_{n-1}(H)$, and let $\sigma \in \pi_n(E)$ denote the class of the canonical cross-section. We shall prove that if $\beta \in \pi_q(Y)$ then

$$(16.4) \quad i_*\langle \alpha, \beta \rangle = -[\sigma, i_*\beta],$$

where $i_* : \pi_*(Y) \to \pi_*(E)$ is induced by the inclusion. In (16.4) the brackets on the left refer to the relative Samelson product and those on the right to the ordinary Whitehead product. Note that (16.4) characterizes the relative Samelson product since the existence of a cross-section implies that i_* is injective.

The general composition law for the relative Samelson product can be obtained as an application. Let $\alpha \in \pi_p(H)$, $\beta \in \pi_q(Y)$, $\gamma \in \pi_r(S^q)$, where $p \geq 1$ and $q, r \geq 2$. The general composition law is given by an expansion of the form

$$(16.5) \quad \langle \alpha, \beta \circ \gamma \rangle = \langle \alpha, \beta \rangle \circ S^p_*\gamma + [\langle \alpha, \beta \rangle, \beta] \circ S^p_*\gamma' + [[\langle \alpha, \beta \rangle, \beta], \beta] \circ S^p_*\gamma'' + \ldots,$$

where $\gamma', \gamma'', \ldots$ are generalized Hopf invariants of γ. This follows at once from (16.4) and the corresponding formula of Barcus and Barratt [13] for Whitehead products. When γ is a suspension the generalized Hopf invariants vanish and the formula reduces to (15.11). When Y is a sphere S^q the quadruple Whitehead products vanish (see [56] or [65]) and the expansion terminates after the third term. Further if q is odd then the triple Whitehead products also vanish (loc. cit.) and the expansion terminates after the second term.

To prove (16.4) we first recall how the Hurewicz isomorphism is defined. Given any space X let $\Omega^q_* X$ $(q = 1, 2, \ldots)$ denote the function-space of free maps of S^q into X. Consider the fibration $\phi : \Omega^q_* X \to X$ defined by evaluation at the basepoint $e \in S^q$. The fibre over the basepoint $e \in X$ is the space $\Omega^q X$ of pointed maps of S^q into X. Choose one such pointed map $u : S^q \to X$ as basepoint in the function-space. Then the adjoint of a pointed map $f' : S^p \to \Omega^q X$ is a map

$f : S^p \times S^q \to X$ and the Hurewicz isomorphism

$$\psi : \pi_p(\Omega^q X) \to \pi_{p+q}(X)$$

is given in terms of the separation element by

$$(16.6) \quad \psi\{f'\} = d(f, u\rho),$$

where $\rho : S^p \times S^q \to S^q$ is as before.

Recall that E, in (16.4), is a bundle over S^n with fibre Y and group H. Choose a map $v : S^{n-1} \to H$ representing the classifying element α, and let f denote the composition

$$S^{n-1} \times Y \xrightarrow{v \times 1} H \times Y \xrightarrow{\mu} Y.$$

If $k : (B^n, S^{n-1}) \to (S^n, e)$ is a relative homeomorphism then, since the induced bundle k^*E is trivial, we can find a relative homeomorphism h, extending f, such that $ph = kl$, as shown below.

$$\begin{array}{ccc} (B^n \times Y, S^{n-1} \times Y) & \xrightarrow{h} & (E, Y) \\ \downarrow l & & \downarrow p \\ (B^n, S^{n-1}) & \xrightarrow{k} & (S^n, e) \end{array}$$

In (16.4), an element $\beta \in \pi_q(Y)$ is given. Choose a representative $u : S^q \to Y$ of β as basepoint in $\Omega^q Y$ and consider the following diagram, where i_* is induced by the inclusion and ψ denotes the Hurewicz isomorphism.

$$\begin{array}{ccccc} \pi_n(\Omega^q_* E, \Omega^q Y) & \xrightarrow{\delta} & \pi_{n-1}(\Omega^q Y) & \xrightarrow{\psi} & \pi_{n+q-1}(Y) \\ \downarrow i_* & & \downarrow i_* & & \downarrow i_* \\ \pi_n(\Omega^q_* E, \Omega^q E) & \xrightarrow[\delta]{} & \pi_{n-1}(\Omega^q E) & \xrightarrow[\psi]{} & \pi_{n+q-1}(E) \end{array}$$

Since $\Omega^q E$ is the fibre of the fibration $\phi : \Omega^q_* E \to E$ we have an isomorphism

$$\phi_* : \pi_n(\Omega_*^q E, \Omega^q E) \approx \pi_n(E).$$

Also G. W. Whitehead's characterization of the Whitehead product, in (3.2) of [152] (see also [160]), shows that

$$\psi\delta(\eta) = -[\phi_*\eta, i_*\beta],$$

for any $\eta \in \pi_n(\Omega_*^q E, \Omega^q E)$. Take $\eta = i_*\zeta$, where $\zeta \in \pi_n(\Omega_*^q E, \Omega^q Y)$ is the class of the adjoint

$$h' : (B^n, S^{n-1}) \to (\Omega_*^q E, \Omega^q Y)$$

of $h(1 \times u)$. Then

$$-[\phi_* i_* \zeta, i_*\beta] = \psi\delta\, i_*(\zeta) = i_*\psi\delta(\zeta),$$

by commutativity of the diagram above. Since $ph = kl$, however, it follows that $\phi_* i_* \zeta = \sigma \in \pi_n(E)$, the class of the canonical cross-section. Also $h|S^{n-1} \times Y$ is given by $f = \mu(v \times 1)$, hence $\delta\zeta$ is represented by the adjoint $f' : S^{n-1} \to \Omega^q Y$ of $f(1 \times u)$, and so

$$\psi\delta(\zeta) = d(f(1 \times u), u\rho) = \langle \alpha, \beta \rangle,$$

by (16.6). Therefore

$$i_*\langle \alpha, \beta \rangle = -[\sigma, i_*\beta],$$

as asserted.

Now suppose that E is a retractible H-bundle over S^n with fibre $Y = G/H$. Choose a retraction $r : E \to Y$. The fibration admits a cross-section of class τ, say, where $\tau \in \pi_n(E)$. Replacing τ by $\sigma = \tau - i_* r_* \tau$ we obtain a class of cross-section such that $r_*\sigma = 0$. Take α, as in (16.4), to be the classifying element of the fibration. If $\beta \in \pi_q(G/H)$ then

$$\langle \alpha, \beta \rangle = r_* i_* \langle \alpha, \beta \rangle = -r_*[\sigma, i_*\beta] = 0,$$

by (16.4) and naturality. Thus the relative Samelson product determines an obstruction to retractibility.

For example, take $E = V_{n+1,k+1}$, with $(G, H) = (O_n, O_{n-k})$ and $Y = V_{n,k}$. Suppose that the fibration is retractible and consider the following diagram.

$$\begin{array}{ccccc} \pi_n(V_{n+1,k+1}) & \xrightarrow{\Delta} & \pi_{n-1}(O_{n-k}) & \xrightarrow{J} & \pi_{2n-k-1}(S^{n-k}) \\ \downarrow p_* & & \downarrow u_* & & \downarrow S^k_* \\ \pi_n(S^n) & \xrightarrow{\Delta} & \pi_{n-1}(O_n) & \xrightarrow{J} & \pi_{2n-1}(S^n) \end{array}$$

The left-hand square is commutative, by naturality, and the right-hand square is commutative apart from sign, as shown in [151] (see also [160]). Write $\gamma = J\Delta\sigma$, where $\sigma \in \pi_n(V_{n+1,k+1})$ is the class of a cross-section. Then

$$\pm S^k_* J\Delta\sigma = Ju_*\Delta\sigma = J\Delta\iota_n = w_n,$$

the Whitehead square. However n is odd, since a cross-section exists, and so $\pm w_n = w_n$. Thus $S^k_*\gamma = w_n$. Also if $\beta \in \pi_{n-k}(V_{n,k})$ denotes the class of the inclusion then

$$\beta \circ \gamma = u_* J\Delta\sigma = \langle \Delta\sigma, \beta \rangle,$$

the obstruction to retractibility, by (16. 2) and naturality. Therefore we obtain

Proposition (16. 7). Suppose that $V_{n,k}$ is a retract of $V_{n+1,k+1}$. Then $\pi_{2n-k-1}(S^{n-k})$ contains an element γ such that

(a) $S^k_*\gamma = w_n$, (b) $\beta \circ \gamma = 0$,

where $\beta \in \pi_{n-k}(V_{n,k})$ denotes the class of the inclusion.

Similar results can be obtained in the complex and quaternionic cases. In §20 below we shall use (16. 7) to prove (1. 11), the triviality theorem.

For another application of the theory take $(G, H) = (R_{q+1}, R_q)$, so that $G/H = S^q$, and consider the operator

$$D : \pi_r(S^q) \to \pi_{r+1}(S^q)$$

defined by taking the relative Samelson product with the generator $\theta \in \pi_1(R_q)$. By (16.2) we have $D\iota_q = J\theta = \eta_q \in \pi_{q+1}(S^q)$. The composition law (16.5) enables us to calculate D in general, as follows. Recall (see [56] or [65]) that the triple Whitehead product in $\pi_*(S^q)$ is of odd order. Also $[\eta_2, \iota_2] = 0$, and $2\eta_q = 0$ for $q > 2$. Hence $[[\eta_q, \iota_q], \iota_q] = 0$ for $q \geq 2$ and so the composition law reduces to

$$(16.8) \quad D\gamma = \eta_q \circ S_*\gamma \pm [\eta_q, \iota_q] \circ S_*H\gamma \quad (\gamma \in \pi_r(S^q)),$$

where H denotes the generalized Hopf invariant. Now $H[\eta_q, \iota_q] = 0$, since $[\eta_q, \iota_q]$ is a suspension element, and $S_*[\eta_q, \iota_q] = 0$. Hence on iterating (16.8) we find, after the first step, that $D^2\gamma = \eta_q \circ \eta_{q+1} \circ S_*^2\gamma$ and then, after two more steps, that $D^4\gamma = 0$, since $\eta_q \circ \eta_{q+1} \circ \eta_{q+2} \circ \eta_{q+3} = 0$. This proves

Proposition (16.9). The operator

$$D^4 : \pi_r(S^q) \to \pi_{r+4}(S^q)$$

is trivial.

By pausing at the third step, incidentally, we see that $D^3\pi_q(S^q) \neq 0$, so that (16.9) is best possible.

With (16.9) in hand let us take a further look at the operator

$$D : \pi_r(R_t) \to \pi_{r+1}(R_t)$$

of §15 and prove

Proposition (16.10). The operator

$$D^6 : \pi_r(R_t) \to \pi_{r+6}(R_t)$$

is trivial whenever t is odd.

First take the case $t \equiv 3 \bmod 4$ and regard D as operating on the fibre homotopy sequence

$$\ldots \to \pi_r(R_{t-1}) \overset{u_*}{\to} \pi_r(R_t) \overset{p_*}{\to} \pi_r(S^{t-1}) \to \ldots$$

as described above. If $\alpha \in \pi_r(R_t)$ then $p_* D^4 \alpha = \pm D^4 p_* \alpha = 0$, by (16. 9) with $q = t - 1$. Hence $D^4 \alpha = u_* \beta$, by exactness, for some $\beta \in \pi_r(R_{t-1})$. But $D^2 \beta = 0$, by (15. 9) with $t - 1$ in place of t, and so

$$D^6 \alpha = D^2 u_* \beta = \pm u_* D^2 \beta = 0.$$

This proves (16. 10) when $t \equiv 3 \bmod 4$. To deal with $t \equiv 1 \bmod 4$ we use the fibration $R_{t+1} \to S^t$, instead of $R_t \to S^{t-1}$, but otherwise the details are similar. The same kind of argument shows that $D^8 = 0$ when $t \equiv 0 \bmod 4$ but this can be improved, as we shall see in a moment.

For a further illustration of the theory take $(G, H) = (R_{2n}, R_{2n-2})$ and hence $G/H = V_{2n, 2}$, where $n \geq 2$. Consider the operator

$$D : \pi_r(V_{2n, 2}) \to \pi_{r+1}(V_{2n, 2})$$

defined by taking the relative Samelson product with the generator $\theta \in \pi_1(R_{2n-2})$. Recall that $J\theta = \eta_{2n-2}$, the generator of $\pi_{2n-1}(S^{2n-2})$. As in §13 let $\alpha_n \in \pi_{2n-2}(V_{2n, 2})$ $(n \geq 2)$ and $\beta_n \in \pi_{2n-1}(V_{2n, 2})$ $(n \geq 1)$ denote the classes of the inclusions

$$S^{2n-2} = V_{2n-1, 1} \overset{u}{\to} V_{2n, 2} \overset{v}{\leftarrow} W_{n, 1} = S^{2n-1}.$$

I assert that

(16. 11) $D\alpha_n = \alpha_n \circ \eta_{2n-2}$, $D\beta_n = \beta_n \circ \eta_{2n-1}$.

For since $\alpha_n = u_* \iota_{2n-2}$ we have

$$\langle \alpha_n, \theta \rangle = u_* \langle \iota_{2n-2}, \theta \rangle = u_* \eta_{2n-2} = \alpha_n \circ \eta_{2n-2},$$

by (16. 2), and similarly

$$\langle \beta_n, \theta \rangle = v_* \langle \iota_{2n-1}, \theta' \rangle = v_* \eta_{2n-1} = \beta_n \circ \eta_{2n-1},$$

where θ' generates $\pi_1(U_{n-1})$. Since any element of $\pi_r(V_{2n, 2})$ $(n \geq 2)$ can be expressed in the form $\alpha_n \circ \sigma + \beta_n \circ \tau$, for some $\sigma \in \pi_r(S^{2n-2})$,

$\tau \in \pi_r(S^{2n-1})$ we can calculate D in general by using (16.11) and the composition law. Proceeding as before we obtain

Proposition (16.12). The homomorphism

$$D^4 : \pi_r(V_{2n,2}) \to \pi_{r+4}(V_{2n,2})$$

is trivial.

Finally let us regard D as operating on the fibre homotopy sequence

$$\ldots \to \pi_r(R_{2n-2}) \xrightarrow{u_*} \pi_r(R_{2n}) \xrightarrow{p_*} \pi_r(V_{2n,2}) \to \ldots .$$

If $\alpha \in \pi_r(R_{2n})$ then $p_* D^4 \alpha = \pm D^4 p_* \alpha = 0$, by (16.12), and so $D^4 \alpha = u_* \beta$, for some $\beta \in \pi_r(R_{2n-2})$. But $D^2 \beta = 0$ when n is even, by (15.9), hence $D^6 \alpha = 0$, which proves

Proposition (16.13). The homomorphism

$$D^6 : \pi_r(R_t) \to \pi_{r+6}(R_t)$$

is trivial for all $t \equiv 0 \bmod 4$.

We have now verified the conjecture (15.8) for all the rotation groups. The corresponding result for the unitary groups makes an interesting exercise.

17 · The Bott suspension

In this section we study the family of maps $B : R_{2n} \to \Omega R_{2n}$ given by the commutator

$$B(x)(t) = [x, e_{2n}\cos \pi t + b_{2n}\sin \pi t] \quad (x \in R_{2n}),$$

where e_{2n} denotes the unit matrix and

$$b_{2n} = \begin{pmatrix} 0 & 1 \\ -1 & 0 \end{pmatrix} \oplus \dots \oplus \begin{pmatrix} 0 & 1 \\ -1 & 0 \end{pmatrix} \quad \text{(n summands).}$$

We call these the <u>Bott maps.</u> Since b_{2n} lies in the centre of U_n there is an induced map $R_{2n}/U_n \to \Omega R_{2n}$; this is essentially the same as the map used in our discussion of relative Stiefel manifolds in §8. We denote by F the homomorphism

$$\pi_r(R_{2n}) \to \pi_r(\Omega R_{2n}) = \pi_{r+1}(R_{2n}) \qquad (r = 1, 2, \dots)$$

induced by the Bott map and refer to this as the (ordinary) <u>Bott suspension.</u> Since $B|U_n$ is constant we have at once that

$$(17.1) \quad Fi_*\pi_r(U_n) = 0,$$

where $i : U_n \subset R_{2n}$. In the stable range the Bott suspension appears (see [90]) in an exact sequence of the form

$$\dots \to \pi_r(U_n) \xrightarrow{i_*} \pi_r(R_{2n}) \xrightarrow{F} \pi_{r+1}(R_{2n}) \to \dots$$

Hence it follows that if $r < 2n$ then

$$(17.2) \quad F\sigma = \sigma \circ \eta \qquad (\sigma \in \pi_r(R_{2n})).$$

It would be interesting to have a formula for F outside the stable range; certainly (17.2) breaks down.

There is another way of defining the Bott suspension, which gives greater insight into its formal properties. Consider the projective group $PR_{2n} = R'_{2n}$ defined by factoring out the central subgroup $\{e, -e\}$. The covering homomorphism $\rho : R_{2n} \to R'_{2n}$ induces the homomorphism

$$\rho_* : \pi_r(R_{2n}) \to \pi_r(R'_{2n}) \qquad (r = 1, 2, \dots)$$

which respects the Samelson product and is an isomorphism for $r \geq 2$.

The situation when $r = 1$ is as follows. The loop $e_{2n}\cos 2\pi t + b_{2n}\sin 2\pi t \quad (0 \leq t \leq 1)$ in R_{2n} represents $n\theta$, where θ generates $\pi_1(R_{2n})$. The loop $\rho(e_{2n}\cos \pi t + b_{2n}\sin \pi t)$ in R'_{2n} represents an element ϕ, say, of $\pi_1(R'_{2n})$. If n is odd then $2\phi = \rho_*\theta$ and ϕ generates $\pi_1(R'_{2n}) = Z_4$. If n is even then $2\phi = 0$ and the pair $(\rho_*\theta, \phi)$ generate $\pi_1(R'_{2n}) = Z_2 \oplus Z_2$. Referring to the definition of the Samelson product in §15 we see at once that

$$(17.3) \quad F\sigma = \rho_*^{-1}\langle \rho_*\sigma, \phi \rangle,$$

for all $\sigma \in \pi_r(R_{2n})$, and of course this can be taken as an alternative definition. If n is odd then $2F = 0$, by linearity, and if n is even then $2F = D$, as in §15.

From the Jacobi identity for the Samelson product in $\pi_*(R'_{2n})$ it follows immediately that F acts as a derivation with respect to the Samelson product in $\pi_*(R_{2n})$.

Now consider the outer automorphism ζ of R_{2n} defined by $g \mapsto aga^{-1}$, where $a = (-e_1) \oplus e_{2n-1} \in O_{2n}$. Let ζ' denote the induced automorphism of R'_{2n}, so that $\rho\zeta = \zeta'\rho$. Clearly $\phi - \zeta'_*\phi = \rho_*\theta$ and so

$$\langle \sigma', \phi \rangle - \langle \sigma', \zeta'_*\phi \rangle = \langle \sigma', \rho_*\theta \rangle,$$

for any element $\sigma' \in \pi_r(R'_{2n})$. However $\langle \sigma', \zeta'_*\phi \rangle = \zeta'_*\langle \zeta'_*\sigma', \phi \rangle$, since ζ' is an involution. If $\sigma' = \rho_*\sigma$, where $\sigma \in \pi_r(R_{2n})$, then $\zeta'_*\sigma' = \rho_*\zeta_*\sigma$ and so we arrive at

$$(17.4) \quad F - \zeta_*F\zeta_* = D.$$

For example, take $n = 2$ with $r = 3$. Then $\zeta_*\alpha = \alpha$, $\zeta_*\beta = \alpha - \beta$, as shown in §23 of [133], where $\alpha, \beta \in \pi_3(R_4)$ denote the standard generators. Now $\beta = i_*\beta'$, where β' generates $\pi_3(U_2)$, and so $\langle \beta', \theta' \rangle = \beta' \circ \eta$, by (16.2), where θ' generates $\pi_1(U_2)$. Hence $\langle \beta, \theta \rangle = \beta \circ \eta$, by naturality, while $F\beta = 0$, by (17.1). Therefore (17.4) shows that

$$(17.5) \quad -F\alpha = \zeta_*(\beta \circ \eta) = \alpha \circ \eta - \beta \circ \eta.$$

Recall moreover, that any element of $\pi_r(R_4)$ can be expressed (uniquely) in the form $\alpha \circ \sigma + \beta \circ \tau$, where $\sigma, \tau \in \pi_r(S^3)$. Hence, using the composition law (16.5), we can now calculate the Bott suspension on $\pi_r(R_4)$, for all values of r.

Returning to the general case, let $1 \leq k < n$. The Bott map

$$B : (R_{2n}, R_{2n-2k}) \to (\Omega R_{2n}, \Omega R_{2n-2k})$$

determines an endomorphism of the homotopy exact sequence of the pair (R_{2n}, R_{2n-2k}), raising dimension by 1. On the absolute groups in the sequence the action is given by the absolute Bott suspension already defined. The action on the relative groups constitutes a homomorphism

$$F : \pi_r(V_{2n,2k}) \to \pi_{r+1}(V_{2n,2k}),$$

called the <u>relative Bott suspension,</u> such that

$$(17.6) \quad Fi_*\pi_r(W_{n,k}) = 0,$$

where $i : W_{n,k} \subset V_{2n,2k}$. Of course the Bott map does not, in general, determine a map $V_{2n,2k} \to \Omega V_{2n,2k}$. However, the relative Bott suspension can be interpreted as a relative Samelson product, as follows.

Let $\overline{R}_{2n-2k} \subset R_{2n}$ denote the direct product of R_{2n-2k} and the circle group generated by $e_{2n-2k} \oplus b_{2k}$. Note that $\overline{R}_{2n-2k}$ contains the circle group generated by b_{2n}. Let $\overline{R}'_{2n-2k} \subset R'_{2n}$ be defined by factoring out the central subgroup $\{e, -e\}$, and let $\rho : R_{2n-2k} \to \overline{R}'_{2n-2k}$ be defined by restriction of the natural projection to R_{2n-2k}. Let $\rho : V_{2n,2k} \to V'_{2n,2k}$ be defined similarly, where $V'_{2n,2k} = R'_{2n}/\overline{R}'_{2n-2k}$. Then ρ maps $V_{2n,2k}$ as an R_{2n-2k}-space into $V'_{2n,2k}$ as an

$\overline{R}'_{2n-2k}$-space, and so the induced homomorphism ρ_* respects the relative Samelson products. Since $\overline{R}_{2n-2k}/R_{2n-2k}$ is a circle, the homomorphism

$$\rho_* : \pi_r(V_{2n,\,2k}) \to \pi_r(V'_{2n,\,2k})$$

is an isomorphism for $r \geq 2$. Let $\overline{\phi} \in \pi_1(\overline{R}'_{2n-2k})$ be defined by the loop $\rho(e_{2n}\cos \pi t + b_{2n}\sin \pi t)$ $(0 \leq t \leq 1)$. It can then be verified that

$$(17.7) \quad F\sigma = \rho_*^{-1}\langle \rho_*\sigma,\ \overline{\phi}\rangle,$$

for all $\sigma \in \pi_r(V_{2n,\,2k})$. Of course this can be taken as an alternative definition of the relative Bott suspension, as in [75].

The original definition makes it clear that F acts as an endomorphism, of degree one, on the homotopy sequence of the fibration

$$V_{2n-2l,\,2k-2l} \to V_{2n,\,2k} \to V_{2n,\,2l},$$

where $1 \leq l < k < n$. With the alternative definition, on the other hand, the relative form of the Jacobi identity (15. 12) for the pair $(R'_{2n},\ \overline{R}'_{2n-2k})$ yields the relation

$$(17.8) \quad F\langle\sigma,\ \tau\rangle = \langle F\sigma,\ \tau\rangle + (-1)^s\langle\sigma,\ F\tau\rangle,$$

for $\sigma \in \pi_r(V_{2n,\,2k})$, $\tau \in \pi_s(R_{2n-2k})$. Thus the Bott suspension acts as a derivation with respect to the Samelson product in the relative sense as well.

Both R_{2n-2k} and $\overline{R}_{2n-2k}$ are stable under the outer automorphism ζ of R_{2n}. Hence it follows by the same argument as we used in the case of (17. 4) that

$$(17.9) \quad F - \zeta_* F \zeta_* = D,$$

where D is as in §15. Taking $k = 1$ we shall apply this, in the next paragraph, to establish the key relation

$$(17.10) \quad F\alpha_n = \alpha_n \circ \eta - \beta_n,$$

where $\alpha_n \in \pi_{2n-2}(V_{2n,\,2})$, $\beta_n \in \pi_{2n-1}(V_{2n,\,2})$ are as before. Notice that

(17.10), by naturality, yields the value of F on the first non-vanishing homotopy group of $V_{2n, 2k}$.

The case $n = 2$ follows from (17.5) by naturality as in (15.13). Let $n > 2$. We have already noted in §7 that $p_* F\alpha_n = \pm\iota_{2n-1}$, where $p : V_{2n, 2} \to S^{2n-1}$. With appropriate conventions, as in [75], the sign turns out to be minus, and so

(17.11) $F\alpha_n = \alpha_n \circ \gamma - \beta_n$,

by exactness, for some element $\gamma \in \pi_{2n-1}(S^{2n-2})$. To determine γ we apply (17.9) to β_n and obtain from (16.11) and (17.6) that

$$\zeta_* F(\alpha_n \circ \gamma) = \beta_n \circ \eta.$$

Now $\zeta_* \beta_n = \mu_* \beta_n = -\beta_n + \alpha_n \circ \eta$, by (13.9), and so

$$F(\alpha_n \circ \eta) = -\zeta_*(\beta_n \circ \eta) = \beta_n \circ \eta - \alpha_n \circ \eta \circ \eta.$$

On the other hand (17.11) yields

$$F(\alpha_n \circ \eta) = \alpha_n \circ \gamma \circ \eta + \beta_n \circ \eta.$$

Therefore $\gamma = \eta$ and (17.10) is established. An alternative proof will be given in the next section.

Any element of $\pi_r(V_{2n, 2})$ can be expressed (uniquely) in the form $\alpha_n \circ \sigma + \beta_n \circ \tau$ where $\sigma \in \pi_r(S^{2n-2})$ and $\tau \in \pi_r(S^{2n-1})$. Now $F(\alpha_n \circ \sigma + \beta_n \circ \tau) = F(\alpha_n \circ \sigma)$, by (17.6), and

$$F(\alpha_n \circ \sigma) = (F\alpha_n) \circ (S_* \sigma) + [\alpha_n, F\alpha_n] \circ (S_* \sigma') + \ldots$$

by the composition law (16.5). Note that $p_*[\alpha_n, F\alpha_n] = 0$, since $p_* \alpha_n = 0$, and so

(17.12) $[\alpha_n, F\alpha_n] = \alpha_n \circ \delta$,

by exactness, for some $\delta \in \pi_{4n-2}(S^{2n-2})$. I assert that

(17.13) $[[\alpha_n, F\alpha_n], \alpha_n] = 0$.

Assuming this, the composition law reduces to

$$(17.14) \quad F(\alpha_n \circ \sigma) = (F\alpha_n) \circ (S_*\sigma) + \alpha_n \circ \delta \circ S_*H\sigma,$$

where $H\sigma \in \pi_r(S^{2n-5})$ denotes the generalized Hopf invariant and δ is as in (17.12).

The proof of (17.13), and hence (17.14), is as follows. By (16.2) and (16.4) (cf. [89]) we have $[\alpha_n, \beta_n] = -\alpha_n \circ J\mu$, where $\mu = \Delta\iota_{2n-1} \in \pi_{2n-2}(R_{2n-2})$. Hence

$$[\alpha_n, F\alpha_n] = [\alpha_n, \alpha_n \circ \eta] - [\alpha_n, \beta_n] = \alpha_n \circ (P\eta + J\mu),$$

by (17.10), where $P\eta = [\eta, \iota_{2n-2}] \in \pi_{4n-4}(S^{2n-2})$. Thus $\delta = P\eta + J\mu$, in (17.12), and so $S_*^2\delta = 0$, since $S_*P\eta = 0$ and

$$S_*^2 J\mu = S_*^2 J\Delta\iota_{2n-1} = S_*w_{2n-1} = 0.$$

Hence $[\delta, \iota_{2n-2}] = \gamma \circ S_*^{2n-2}H\delta$, by the composition law of [13] for Whitehead products, where $\gamma = [\iota_{2n-2}, [\iota_{2n-2}, \iota_{2n-2}]]$. But $3\gamma = 0$, by the Jacobi identity, while $H\delta$ is a multiple of η. Since $2\eta = 0$ this implies that $[\delta, \iota_{2n-2}] = 0$, hence $[\alpha_n, \alpha_n \circ \delta] = 0$, by naturality, which proves (17.13). Thus we can now calculate the action of the Bott suspension on $\pi_r(V_{2n,2})$, for all values of r.

For example, it follows by repeated application of (17.14) that the iterated Bott suspension

$$F^6 : \pi_r(V_{2n,2}) \to \pi_{r+6}(V_{2n,2})$$

is always trivial. Hence, by induction on k, it follows that

$$F^{6k} : \pi_r(V_{2n,2k}) \to \pi_{r+6k}(V_{2n,2k})$$

is always trivial. In particular

$$F^{6n} : \pi_r(R_{2n}) \to \pi_{r+6n}(R_{2n})$$

is trivial. Stronger results of this type can be found in §5 of [75].

The Bott suspension is also useful for computing relative Samelson products, such as

$$\langle \sigma, \beta_n \rangle \in \pi_{r+2n-2}(V_{2n,2}) \qquad (\sigma \in \pi_r(R_{2n-2})).$$

By (16.2) we have at once that

$$(17.15) \quad \langle \sigma, \beta_n \rangle = \alpha_n \circ \tau + \beta_n \circ S_* J\sigma,$$

for some $\tau \in \pi_r(S^{2n-2})$, and therefore

$$\langle F\sigma, \beta_n \rangle = F\langle \sigma, \beta_n \rangle = F\langle \alpha_n \circ \tau),$$

by (17.6) and (17.8). However

$$F(\alpha_n \circ \tau) = \alpha_n \circ \tau' + \beta_n \circ S_* \tau$$

by (17.10), (17.14), where $\tau' \in \pi_r(S^{2n-2})$, while

$$\langle F\sigma, \beta_n \rangle = \alpha_n \circ \tau'' + \beta_n \circ S_* JF\sigma,$$

by (16.2), where $\tau'' \in \pi_r(S^{2n-2})$. Thus

$$(17.16) \quad S_* \tau = S_* JF\sigma.$$

Substituting $F\sigma$ for σ, in this relation, we obtain $S_* \tau'' = S_* JF^2 \sigma$, while $\tau' = \eta \circ S_* \tau + \delta \circ S_* H\tau$, by (17.10) and (17.14). Since $\tau' = \tau''$ and $S_*^2 \delta = 0$ this shows that

$$(17.17) \quad S_*^2 JF^2 \sigma = S_*^2 (\eta \circ S_* JF\sigma).$$

If S_*^2 is injective, as when $r < 4n - 5$, this implies that $JF^2 \sigma = \eta \circ S_* JF\sigma$; moreover $\tau = JF\sigma$, by (17.16), and so

$$\langle \sigma, \beta_n \rangle = \alpha_n \circ JF\sigma + \beta_n \circ S_* J\sigma.$$

18 · The intrinsic join again

The main purpose of this section is to deform the intrinsic map of §2 into a map which can be expressed in terms of commutators. The deformation, which is due to Husseini [61], establishes a conjecture of Bott [30]. Various relations between the intrinsic join and the Samelson product are deduced.

Recall that a_r $(r = 1, 2, \ldots)$ denotes the rth basis vector in s-space, for $s \geq r$, and that our standard embedding $u : G_t \to G_s$ $(t \leq s)$ is that which leaves the last $s - t$ basis vectors fixed. Let $u' : G_t \to G_s$ be the embedding which leaves the first $s - t$ basis vectors fixed. In G_{m+n}, therefore, the subgroups $G_m = uG_m$ and $G'_n = u'G_n$ commute. Given k, where $k \leq m, n$, let d_t denote simultaneous rotation through $\frac{1}{2}\pi t$ in each of the planes

$$(a_{m-k+1}, a_{m+n-k+1}), \ldots, (a_m, a_{m+n}).$$

Thus d_t is given by the matrix shown below where $c = \cos \frac{1}{2}\pi t$, $s = \sin \frac{1}{2}\pi t$.

e_{m-k}			
	ce_k		se_k
		e_{n-k}	
	$-se_k$		ce_k

In the real case when m, n and k are even we note, for future reference, that d_t is given by a unitary transformation.

Let $H : G_m \times G_n \times I \to G_{m+n}$ denote the map defined by

$$H(x, y, t) = [d_t y' d_t^{-1}, x]$$

where $x \in G_m$, $y \in G_n$ and $y' = u'y \in G_{m+n}$. The commutator depends only on the coset of x in G_m/G_{m-k}, and on the coset of y in G_n/G_{n-k}. Also $H(x, y, 0)$ is trivial and $H(x, y, 1) \in G_{m+n-k}$, for all x, y, and so H induces a map

$$g : S(O_{m,k} \wedge O_{n,k}) \to O_{m+n,k}.$$

Recall that the natural projection

$$\chi : O_{m,k} * O_{n,k} \to S(O_{m,k} \wedge O_{n,k})$$

is a homotopy equivalence. Bott conjectured that the composition

$$h' = g\chi : O_{m,k} * O_{n,k} \to O_{m+n,k}$$

was essentially the same as the intrinsic map h of §2. We shall prove, following Husseini, that

$$(18.1) \quad h' \simeq h(T * 1),$$

where T denotes the self-map of $O_{m,k}$ which changes the sign of all the vectors in each k-frame. There is also a relation between g and the commutator map

$$c : G_m \wedge G_n \to G_{m+n-k},$$

given by $c(x, y) = [x, y] = [y, x]^{-1}$, where $x \in G_m$, $y \in G_n$. Consider the transgression operator

$$\Delta : \pi(S(G_m \wedge G_n), G_{m+n,k}) \to \pi(G_m \wedge G_n, G_{m+n-k})$$

in the homotopy exact sequence of the fibration

$$G_{m+n-k} \to G_{m+n} \to O_{m+n,k}.$$

After referring back to the definition of h we see that

$$(18.2) \quad \Delta\{g(Sp')\} = \{c^{-1}\},$$

where $p' : G_m \wedge G_n \to G_{m,k} \wedge G_{n,k}$ denotes the smash product of the projections.

To prove (18.1), consider the deformation

$$H_s : G_m \times G_n \times I \to G_{m+n} \qquad (0 \le s \le 1)$$

of H which transforms (x, y, t) into

$$d_{(1-2)t} y'' d^{-1}_{(1-s)t} x d_t y''^{-1} d_{-s+(1-s)t} x^{-1},$$

where $y'' = d_1 y d_1^{-1}$. Recall that the projection $p = p_{m+n} : G_{m+n} \to O_{m+n,k}$ is given by evaluation at the point v_0, say, where

$$v_0 = (a_{m+n-k+1}, \dots, a_{m+n})$$

is left fixed by G_{m+n-k}. Now pH_s depends only on the cosets $p_m x \in O_{m,k}$, $p_n y \in O_{n,k}$. Moreover pH_s is independent of x when $t = 1$ and of y when $t = 0$. Therefore pH_s induces a homotopy

$$K_s : O_{m,k} * O_{n,k} \to O_{m+n,k},$$

such that $K_0 = g\chi = h'$. To complete the proof of (18.1) we check that $K_1 = h(T * 1)$. Write $u_0 = -d_1 v_0 = (-a_{m-k+1}, \dots, -a_m)$, so that $xv_0 = v_0$ while $y''u_0 = u_0$. Then $p_m x = xu_0$, $p_n y = yv_0$, where $x \in G_m$, $y \in G_n$. Also

$$\begin{aligned} K_1(xu_0, yv_0, t) &= y''xd_t(y''^{-1}d_1^{-1}v_0) \\ &= y''xd_t u_0 \\ &= (-xu_0 \cos \pi t, \ yv_0 \sin \pi t). \end{aligned}$$

Thus $K_1 = h(T * 1)$ and the proof of (18.1) is complete.

From (18.1) and (18.2) we obtain the useful relation

$$(18.3) \quad \Delta\mu^m_*(p_*\alpha * p_*\beta) = -\langle u_*\alpha, u_*\beta\rangle,$$

for $\alpha \in \pi_i(G_m)$, $\beta \in \pi_j(G_n)$, as shown in the following diagram.

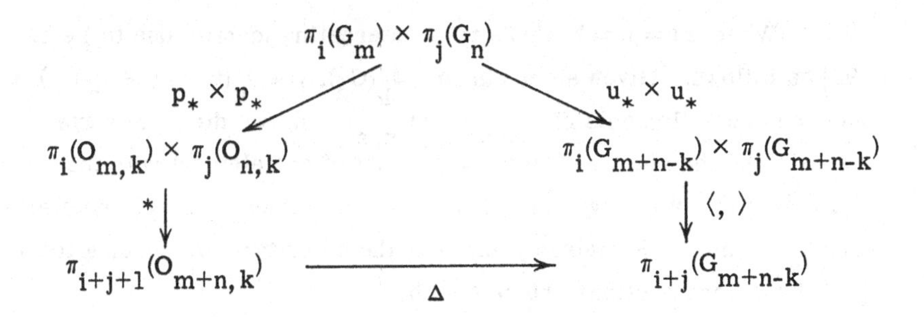

The remainder of this section is devoted to establishing some further relations between the intrinsic join and the Samelson product. For this purpose it is convenient to introduce another pairing

$$\pi_i(G_m, G_{m-k}) \times \pi_j(G_n, G_{n-k}) \to \pi_{i+j+1}(G_{m+n}, G_{m+n-k}),$$

defined as follows. Consider the map

$$H^{-1} : G_m \times G_n \times I \to G_{m+n},$$

where H is as before, so that

$$H^{-1}(x, y, t) = [x, d_t y' d_t^{-1}].$$

We see at once, from our previous remarks, that H^{-1} induces a map

$$L : (G_m * G_n, G_m * G_{n-k} \cup G_{m-k} * G_n) \to (G_{m+n}, G_{m+n-k}).$$

The ordinary relative join is a pairing of $\pi_i(G_m, G_{m-k})$ with $\pi_j(G_n, G_{n-k})$ to

$$\pi_{i+j+1}(G_m * G_n, G_m * G_{n-k} \cup G_{m-k} * G_n).$$

The new pairing, which we denote by $*'$, is defined by applying the induced homomorphism L_* to the relative join. If we identify $\pi_*(G_r, G_{r-k})$ with $\pi_*(O_{r,k})$ in the usual way, for $r = m, n, m+n$, we obtain from (18.1) that

$$(18.4) \quad \theta *' \phi = -(\mu_*)^m(\theta * \phi),$$

where $\theta \in \pi_i(O_{m,k})$, $\phi \in \pi_j(O_{n,k})$.

When $m = n = k$ there is an interesting identity due to Leise [95] as follows. Given elements $\alpha \in \pi_p(G_n)$, $\beta \in \pi_q(G_n)$, $\gamma \in \pi_r(G_n)$ we can construct elements of $\pi_{p+q+r}(O_{2n,n})$ in two different ways: we can take the relative Samelson product of one element with the intrinsic join of the other two, or we can take the intrinsic join of one element with the ordinary Samelson product of the other two. According to Leise [95] these constructions are related by

$$(18.5) \quad \alpha * \langle\beta, \gamma\rangle = (-1)^{p+q+r}\langle\alpha * \beta, \gamma\rangle + (-1)^{pq+p+r+1}\langle\beta, \alpha * \gamma\rangle.$$

To prove this, notice first that d_t in the present case reduces to the matrix

$$\begin{pmatrix} ce_n & se_n \\ -se_n & ce_n \end{pmatrix}$$

where $c = \cos\frac{1}{2}\pi t$, $s = \sin\frac{1}{2}\pi t$, as before. Now $z' = d_1 z d_1^{-1}$, for any $z \in G_n$, hence $d_t z' d_t^{-1} = d_{1+t} z d_{1+t}^{-1}$. Represent the given elements α, β, γ by maps f, g, h, where $f : (I^p, \dot{I}^p) \to (G_n, e)$, etc; and write $f_s(x) = d_{1+s}(fx)d_{1+s}^{-1}$, where $x \in I^p$, $s \in I$. Consider the homotopies

$$k_t, l_t, m_t : (I^{p+q+r+1}, \dot{I}^{p+q+r+1}) \to (G_{2n}, G_n)$$

which are given on $x \in I^p$, $y \in I^q$, $z \in I^r$ and $s, t \in I$ by

$$k_t(x, y, z, s) = [h(tz)f_s(x)h^{-1}(tz), [gy, hz]]$$

$$l_t(x, y, z, s) = [f_s(tx)g(y)f_s^{-1}(tx), [hz, f_s x]]$$

$$m_t(x, y, z, s) = [g(ty)h(z)g^{-1}(ty), [f_s x, gy]].$$

The product $k_t \cdot l_t \cdot m_t$ is trivial when $t = 1$, by (15.1). Taking $t = 0$ it is easy to check that l_0 represents

$$(-1)^{p(q+r)}\langle\beta, \gamma *' \alpha\rangle,$$

that m_0 represents

$$-(-1)^{qr+pr+pq}\langle\gamma, \beta *' \alpha\rangle$$

and that k_0 represents

$$-(-1)^{pq+pr}\langle\beta, \gamma\rangle *' \alpha.$$

Hence the sum of these elements is zero and by using (18. 4) and the commutation law we arrive at (18. 5).

In §2 of [75] it is stated, without proof, that the Bott suspension acts as a derivation with respect to the intrinsic join. Although I have no reason to doubt the truth of this assertion I have not been able to find a short and convincing proof. Let us leave this aside as a conjecture, therefore, and content ourselves by establishing a useful special case, namely

Theorem (18. 6). <u>If</u> $\alpha \in \pi_p(V_{2m, 2k})$ <u>and</u> $\beta \in i_*\pi_q(W_{n,k}) \subset \pi_q(V_{2n, 2k})$ <u>then</u>

$$F(\alpha * \beta) = (F\alpha) * \beta.$$

Recall from §6 and §17 that the Bott suspension is defined through conjugation by $b_t = e_{2n}\cos \pi t + b_{2n}\sin \pi t$, which lies in the centre of U_n. Also recall from earlier in this section that d_t is a unitary transformation in this case. To prove (18. 6) we represent α, β by maps

$$f : (I^p, \dot{I}^p) \to (O_{2m}, O_{2m-2k}), \quad g : (I^q, \dot{I}^q) \to (U_n, U_{n-k}).$$

Consider the homotopy

$$\xi_u : (I^{p+q+2}, \dot{I}^{p+q+2}) \to (O_{2m+2n}, O_{2m+2n-2k})$$

defined for $x \in I^p$, $y \in I^q$, s, t, $u \in I$ by

$$\xi_u(x, y, s, t) = [f(ux)b_t(f(ux))^{-1}, [fx, g_s y]],$$

where $g_s(y) = d_s g(y) d_s^{-1}$. Clearly ξ_0 represents $F(\alpha *' \beta)$. Now $g_s(y) \in U_n$ and so commutes with b_t. Hence it follows from (15. 1) that $\xi_1 \cdot h''$ is constant, where

$$h''(x, y, s, t) = [g_s y, [b_t, fx]].$$

But h" represents $-\beta *' F\alpha$ and so, using (18.4), we obtain (18.6).

As an application we give an alternative proof of the key formula (17.10) for $F\alpha_n$. The case $n = 2$ is deduced from (17.6), as before. Omitting $n = 3$, for the moment, let $n \geq 4$. Then

$$\alpha_n = \alpha_2 * \beta_{n-2}, \quad \beta_n = \beta_2 * \beta_{n-2}.$$

Since β_{n-2} is represented by $W_{n-2,1} \subset V_{2n-4,2}$ we obtain

$$\begin{aligned} F\alpha_n &= F\alpha_2 * \beta_{n-2}, \text{ by (18.6)}, \\ &= (\alpha_2 \circ \eta - \beta_2) * \beta_{n-2} \\ &= \alpha_n \circ \eta - \beta_n, \end{aligned}$$

as required. This shows, in particular, that

$$F\alpha_3 * \beta_2 = F\beta_5 = \alpha_5 \circ \eta - \beta_5 = (\alpha_3 \circ \eta - \beta_3) * \beta_2,$$

hence $F\alpha_3 = \alpha_3 \circ \eta - \beta_3$, by the generalized Freudenthal theorem. This completes the proof.

19 · Homotopy-commutativity

Let G', G'' be subgroups of a topological group G. The Samelson product in $\pi_*(G)$ determines a pairing of $\pi_i(G')$ with $\pi_j(G'')$ to $\pi_{i+j}(G)$. We call this the Samelson pairing and continue to use the $\langle , \rangle$ notation. Of course the pairing can be defined directly through the commutator $c : G' \wedge G'' \to G$. If c is nulhomotopic we say that G' homotopy-commutes with G'' in G. This cannot happen unless the Samelson pairing vanishes.

Since G_n is conjugate, in G_{m+n}, to a subgroup G'_n which commutes with G_m, it follows at once that G_m and G_n homotopy-commute in G_{m+n}, for all values of m and n. The purpose of this section is to investigate, in each of the three cases, whether this result can be improved. We follow James and Thomas [86] in the real case and Bott [30] in the others.

Let us begin with the unitary case and consider the exact sequence

$$\pi_{2t+1}(U_{t+1}) \xrightarrow{p_*} \pi_{2t+1}(S^{2t+1}) \xrightarrow{\Delta} \pi_{2t}(U_t) \to 0.$$

For $t \geq 1$ it has been shown by Bott [27] that coker $p_* \approx \pi_{2t}(U_t)$ is cyclic of order $t!$, and this holds for $t = 0$ under the convention $0! = 1$. Take $t = m + n - 1$, where $m, n \geq 1$. Choose $\alpha \in \pi_{2m-1}(U_m)$ so that $p_*\alpha$ generates the image of p_*, and choose $\beta \in \pi_{2n-1}(U_n)$ similarly. Then it follows immediately from (18.3) that

(19.1) $\langle \alpha, \beta \rangle = (m-1)!\,(n-1)!\,\gamma$

where γ generates $\pi_{2t}(U_t)$. Thus $\langle \alpha, \beta \rangle \neq 0$ unless $m = n = 1$, and we obtain

Proposition (19.2). Let $m, n \geq 1$. Then U_m and U_n homotopy-commute in U_{m+n-1} if and only if $m = n = 1$.

Turning to the symplectic case, we recall from [27] that $\pi_{4t+2}(Sp_t)$ is cyclic of order $(2t-1)!$ or $(2t-1)!2$ according as t is odd or even. Again the results of the previous section are used to compute the Samelson pairing and we obtain

Proposition (19.3). <u>Let</u> $m, n \geq 1$. <u>Then</u> Sp_m <u>and</u> Sp_n <u>do not homotopy-commute in</u> Sp_{m+n-1}.

In the real case the problem is more subtle, and has not yet been completely solved. Let us describe the pair (m, n) as <u>irregular</u> if O_m and O_n homotopy-commute in O_{m+n-1}, as <u>regular</u> if they do not. The pair (m, n) is irregular whenever $m + n = 4$ or 8, since then O_{m+n-1} is a retract of O_{m+n}. Are there any other irregular pairs? We shall prove

Theorem (19.4). <u>Let</u> $m + n \neq 4$ <u>or</u> 8. <u>Then</u> (m, n) <u>is regular if</u> m <u>or</u> n <u>is even or if</u> $d(m) = d(n)$, <u>where</u> $d(q)$, <u>for</u> $q \geq 2$, <u>denotes the greatest power of two which divides</u> $q - 1$.

Thus the answer to our question is negative for $m + n < 12$. The least pairs for which the answer is unknown are $(3, 9)$ and $(5, 7)$. If $m > 2$ then either (m, n) or $(m, n-1)$ is regular, by (19.4), and so we deduce

Corollary (19.5). <u>If</u> $m + n > 4$ <u>then</u> O_m <u>and</u> O_n <u>do not homotopy-commute in</u> O_{m+n-2}.

This shows that the least value of q such that O_m and O_n homotopy-commute in O_q is $q = m + n$ when (m, n) is regular and $q = m + n - 1$ otherwise. Another consequence is

Corollary (19.6). <u>The pair</u> (m, m) <u>is irregular if and only if</u> $m = 2$ <u>or</u> 4.

Before we begin the proof of (19.4), there is another problem which perhaps deserves a brief mention. Let us say that a subgroup H of a topological group G is <u>homotopy-normal</u> if the commutator map $G \wedge H \to G$ can be deformed into H. Excluding the real case when $k = 1$ it can be shown, as in [74], that G_k is not homotopy-normal in

G_n, for $1 \leq k < n$.

Returning to the problem of homotopy-commutativity, we base the proof of (19.4) on

Lemma (19.7). *If* (m, n) *is an irregular pair then there exists an* $(m + n)$*-plane bundle* V *over* $S(P^{m-1} * P^{n-1})$ *for which the* mod 2 *Stiefel-Whitney class* w_{m+n} *is non-zero.*

The fibration $p : O_{q+1} \to S^q$ maps P^q with degree 1, in mod 2 cohomology. The induced homomorphism p^* carries the generator $c_q \in H^q(S^q)$ into an element $u_q \in H^q(O_{q+1})$ which transgresses into the Stiefel-Whitney class w_{q+1} of the universal $(q + 1)$-plane bundle. For this bundle the transgression determines a $(1, 1)$ correspondence, for each complex L, between equivalence classes of $(q + 1)$-plane bundles over the suspension SL and homotopy classes of maps $f : L \to O_{q+1}$. Further the $(q + 1)$-st Stiefel-Whitney class of the bundle corresponding to f is the suspension $S^*f^*u_q \in H^{q+1}(SL)$ of $f^*u_q \in H^q(L)$.

Now take $L = P^{m-1} * P^{n-1}$ and suppose that $g = pf$, for some f, where $g : P^{m-1} * P^{n-1} \to S^{m+n-1}$ is the map of degree one obtained from

$$h = p * p : O_m * O_n \to S^{m-1} * S^{n-1} = S^{m+n-1}$$

by restriction. Then

$$f^*u_{m+n-1} = f^*p^*c_{m+n-1} = g^*c_{m+n-1} \neq 0,$$

and so $w_{m+n} \neq 0$ on the bundle corresponding to f. Certainly g can be lifted to O_{m+n} if h can and so, from (18.2), with $k = 1$, we obtain (19.7).

Products are trivial in the cohomology ring of a suspension SL. For a real vector bundle over SL, therefore, the formula of Wu [165] for the Steenrod squares of the Stiefel-Whitney classes reduces to

$$(19.8) \quad Sq^t w_r = \binom{r-1}{t} w_{r+t}.$$

If L itself is a suspension then the Pontrjagin square acts trivially in the cohomology of SL and so

$$(19.9)\quad \rho_4(p_r) = w'_{4r},$$

by the second formula of Wu [166]. Here w'_{4r} denotes the image of w_{4r} under the coefficient monomorphism $Z_2 \to Z_4$, and $\rho_4(p_r)$ denotes the mod 4 reduction of the Pontrjagin class p_r. We now use these results to prove

Lemma (19.10). *Let* m *and* n *be even, let* $m + n = 0 \bmod 4$, *and let* $m + n \geq 16$. *Let* V *be a real vector bundle over* $S(P^{m-1} * P^{n-1})$. *Then the Stiefel-Whitney class* w_{m+n} *of* V *is zero.*

By suspending the join of the coverings

$$S^{m-1} \to P^{m-1}, \quad S^{n-1} \to P^{n-1}$$

we obtain a map $S^{m+n} \to SL$ of degree 4. Write $m + n = 4r$. Bott [27] has shown that the Pontrjagin class p_r of a bundle over S^{4r} is always divisible by $(2r - 1)!$. We have $r \geq 4$ and so the mod 4 reduction of p_r is trivial in the case of V. The base of V is a double suspension and so (19.9) applies. Therefore $w'_{m+n} = 0$, and since there is no torsion in dimension $m + n$ it follows that $w_{m+n} = 0$, as asserted. In view of (19.7) we deduce that pairs (m, n) are regular if they satisfy the hypothesis of (19.10).

The proof of (19.4) is completed by application of (19.8), as follows. A basis for the mod 2 cohomology of the space

$$S(P^{m-1} * P^{n-1}) = S^2(P^{m-1} \wedge P^{n-1})$$

consists of the elements

$$\sigma^2(x^i \otimes y^j) \quad (0 < i < m, \quad 0 < j < n),$$

where x generates $H^1(P^{m-1})$, y generates $H^1(P^{n-1})$ and $\sigma = S^*$. Let E denote the subspace spanned by these basis elements where i and j are even. Then $Sq^t E \subset E$, by the Cartan formula, and in particular $Sq^1 E = 0$. We prove

Lemma (19.11). *Let* $u \in H^r(S(P^{m-1} * P^{n-1}))$, *where* r *is even and* $r < m + n$. *If* $Sq^1 u = 0$ *then* $u \in E$.

We have $u \equiv v \bmod E$ for some element v of the form

$$v = \Sigma a_{ij} \sigma^2 (x^i \otimes y^j) \qquad (a_{ij} \in Z_2),$$

with summation confined to odd values of i and j. It follows from the Cartan product formula that

$$Sq^1 v = \Sigma a_{ij} \sigma^2 (x^{i+1} \otimes y^j + x^i \otimes y^{j+1}),$$

since i and j are both odd. By hypothesis $Sq^1 u = 0$ and so $Sq^1 v = 0$ since $Sq^1 E = 0$. Also $i + j + 2 < m + n$ and so either $i + 1 < m$ or $j + 1 < n$. Hence all the coefficients a_{ij} are zero. Thus $v = 0$ and $u \in E$, as asserted.

Now consider the Stiefel-Whitney classes $w_1, w_2, \ldots$ of any vector bundle V over $S(P^{m-1} * P^{n-1})$. We prove

Lemma (19.12). Let $r \leq m + n$ and, if $m + n$ is a power of 2, let $r < m + n$. Then $w_r \in E$.

Since $S(P^{m-1} * P^{n-1})$ is 3-connected, the lemma is trivial for $r \leq 3$. Let $r \geq 4$ and write $r = s + t$, where s is a power of 2 and $0 \leq t < s$. By (19.8) we have that

$$Sq^k w_{k+1} = w_{2k+1} \qquad (k = 1, 2, 4, \ldots, s/2).$$

Hence $w_{s+1} = 0$, by induction, since $w_2 = 0$. But $w_{s+1} = Sq^1 w_s$ by (19.8), since s is even, and $w_s \in E$, by (19.11). However $Sq^t E \subset E$ and $Sq^t w_s = w_r$, by (19.8) again, which proves (19.12). Finally we prove

Lemma (19.13). Suppose that $m = as + 1$ and $n = bs + 1$, where a, b are odd and $s \geq 2$ is a power of 2. Then $w_{m+n} = 0$.

For consider the operator

$$\alpha = Sq^{st} \circ \ldots \circ Sq^{8t} \circ Sq^{4t} \circ Sq^{2t},$$

where $2t = a + b$. If $i + j = a + b$ the Cartan product formula yields

$$\alpha(x^i \otimes y^j) = x^{si} \otimes y^{sj},$$

which is zero unless $i = a$ and $j = b$. Since a and b are odd this shows that $\alpha u = 0$ if $u \in E$ and $\dim u = a + b + 2$. However $w_{m+n} = \alpha w_{a+b+2}$, by (19. 8), since $a + b$ is even; and $w_{a+b+2} \in E$, by (19. 12). Therefore $w_{m+n} = 0$, as asserted.

To obtain (19. 4) we combine (19. 7) with the various lemmas we have proved. We use (19. 12) when $m + n$ is odd, also when $m + n$ is not a power of 2 and both m and n are even. We use (19. 13) when both m and n are odd. The remaining cases are covered by (19. 10) and so (19. 4) is established.

20 · The triviality problem

Throughout this section we use cohomology with mod 2 coefficients. Recall (see [6]) that the algebra of Steenrod squaring operations is generated by the operations Sq^1, Sq^2, Sq^4, Sq^8, ... This result has been refined by Adams [1] as follows. Let K be a complex and let $s = 2^{r+1}$ for some $r \geq 0$. We describe an element $u \in H^n(K)$ as a Sq^s-*class* if $Sq^t u = 0$ for $t = 1, 2, \ldots, s - 1$, while $Sq^s u \neq 0$. Suppose that u satisfies this condition for some $r \geq 3$. Adams shows that there exist cohomology classes u_{ij} $(1 \leq i \leq j \leq r;\ i \neq j - 1)$, of dimension $n + 2^i + 2^j - 1$, such that

$$(20.1) \quad Sq^s u = \Sigma a_{ij} u_{ij} \qquad (s = 2^{r+1})$$

where the a_{ij} are elements of the Steenrod algebra having degree $2^{r+1} + 1 - 2^i - s^j$. Since a_{rr} is of degree one it follows that either $Sq^1 H^{n+s-1}(K) \neq 0$ or $H^{n+t}(K) \neq 0$ for some $t = 2^i + 2^j - 1$ $(1 \leq i \leq j \leq r;\ i \neq j - 1;\ i \neq r)$. For our purposes, however, it is the dual of Adams' result which is more convenient, and we state this as

Theorem (20.2). *Let* $u \in H^n(K)$ *be a* Sq^s*-class, where* $s = 2^{r+1}$ *and* $r \geq 3$. *Then either* $Sq^1 H^n(K) \neq 0$ *or* $H^{n+s-t}(K) \neq 0$ *for some* $t = 2^i + 2^j - 1$ $(1 \leq i \leq j \leq r;\ i \neq j - 1;\ i \neq r)$.

If the Whitehead square $w_n \in \pi_{2n-1}(S^n)$ vanishes, there exists a map $S^n \times S^n \to S^n$ of type (1, 1), hence a map $S^{2n+1} \to S^{n+1}$ of Hopf invariant one. The mapping cone of the latter has non-trivial Sq^{n+1}. Hence Adem [6] showed that $n + 1$ is a power of two and then Adams [1] that $n = 1$, 3 or 7, using the machinery described above. The argument which follows is based on the same procedure.

Let V be an $(n - k - 1)$-connected complex, where $n > k \geq 1$, and let $\beta \in \pi_{n-k}(V)$ be an element such that the induced homomorphism

$$\beta^* : H^{n-k}(V) \to H^{n-k}(S^{n-k})$$

is non-trivial. Suppose that there exists an element $\gamma \in \pi_{2n-k-1}(S^{n-k})$ such that

$$(20.3) \quad \begin{cases} \text{(a)} & \beta \circ \gamma = 0, \\ \text{(b)} & S_*^k \gamma = w_n. \end{cases}$$

Write $\beta' = S_*^k \beta \in \pi_n(S^k V)$. Then

$$[\beta', \beta'] = \beta'_* w_n = S_*^k(\beta \circ \gamma) = 0,$$

by (20.3), and so there exists a map

$$\theta : S^n \times S^n \to S^k V$$

of type (β', β'). Consider the mapping cone $L = e^{2n+2} \cup S^{k+1} V$ of the map

$$\phi : S^{2n+1} \to S^{k+1} V$$

obtained from θ by the Hopf construction. In the cohomology exact sequence

$$H^r(L, S^{k+1}V) \xrightarrow{i^*} H^r(L) \xrightarrow{j^*} H^r(S^{k+1}V),$$

let $\lambda \in H^{2n+2}(L)$ denote the image under i^* of a generator of the relative group. Now $\beta^*\sigma \neq 0$, by hypothesis, for some element $\sigma \in H^{n-k}(V)$, and so $\beta'^*\sigma' \neq 0$, where $\sigma' \in H^n(S^k V)$ is the k-fold suspension of σ. A straightforward calculation, as in (1.4) of [141], shows that

$$(20.4) \quad \lambda = \mu \cup \mu = Sq^{n+1}\mu,$$

where $\mu \in H^{n+1}(L)$ is the element such that $j^*\mu = S^*\sigma$. We need a condition to ensure that $\lambda \neq 0$, such as

Lemma (20.5). <u>Suppose that</u> $H^{2n-k}(V)$ <u>is spanned by decomposable elements. Then</u> L <u>can be constructed, by suitable choice of</u>

θ, so that $\lambda \neq 0$.

For let $T = e^{2n-k} \cup S^{n-k}$ denote the mapping cone of γ. By (20.3a) there exists a map $t : T \to V$ such that $t|S^{n-k}$ represents β. Now T has trivial cup-products since $n > n - k$. Hence

$$t^* : H^{2n-k}(V) \to H^{2n-k}(T)$$

is trivial, and hence so is

$$(S^k t)^* : H^{2n}(S^k V) \to H^{2n}(S^k T).$$

By (20.3b) the 2n-cell of $S^k T = e^{2n} \cup S^n$ is attached by the Whitehead square w_n. We can therefore regard $S^k T$ as obtained from $S^n \times S^n$ by identifying axes in the usual way. Choose $\theta = S^k t \circ \psi$, where

$$\psi : S^n \times S^n \to S^k T$$

is the identification map. Then

$$\theta^* : H^{2n}(S^k V) \to H^{2n}(S^n \times S^n)$$

is trivial, and hence so is

$$\phi^* : H^{2n+1}(S^{k+1} V) \to H^{2n+1}(S^{2n+1}).$$

Since L is the mapping cone of ϕ this implies that $\lambda \neq 0$, as asserted.

Now suppose that V contains a subspace P such that P is an S-retract of V. Also suppose that $\beta \in l_* \pi_{n-k}(P)$, where $l : P \subset V$. Choose $r > 2n - k$ such that a retraction $\rho : S^r V \to S^r P$ exists, and consider the complex

$$e^{r+2n-k+1} \cup S^r P$$

obtained from $S^{r+2n-k+1} L$ by identifying points of $S^r V$ with their images under ρ. With all these hypotheses, including (20.3), we see that (20.4) and (20.5) imply

Proposition (20.6). _Suppose that $H^{2n-k}(V)$ is spanned by decomposable elements. Then for some integer r there exists a com-_

plex K of the form

$$e^{r+2n-k+1} \cup S^r P,$$

such that $Sq^{n+1} H^{r+n-k}(K) \neq 0$.

We apply this to prove (1.11), with n and k increased by one. By hypothesis $V_{n+1,k+1}$ is trivial, as a fibre space over S^n, hence retractible. By (16.7) there exists an element $\gamma \in \pi_{2n-k-1}(S^{n-k})$ such that $S^k_* \gamma = w_n$ and such that $\beta \circ \gamma = 0$, where $\beta \in \pi_{n-k}(V_{n,k})$ is the class of the inclusion. Moreover decomposable elements generate $H^{2n-k}(V_{n,k})$, by (3.8). In view of (7.10) we can apply (20.6) to the pair $(V_{n,k}, P_{n,k})$ and obtain a complex K of the form $e^{r+2n-k+1} \cup S^r P_{n,k}$ such that $Sq^{n+1} H^{r+n-k}(K) \neq 0$. Since $H^{r+s}(K) = 0$ for $s = n, \ldots, 2n-k$ the Adem relations imply that the interval $[n - k + 1, \ldots, n + 1]$ contains a power of two. But $n + 1 \equiv 0 \mod 2^{\sigma(k+1)}$, since the fibration admits a cross-section, and so the interval $[n - k + 1, \ldots, n]$ contains no power of two. Therefore $n + 1$ is a power of two, which proves the first part of (1.11).

In the second part k is odd, hence $n - k$ is even, and so $Sq^1 H^{n-k+r}(K) = 0$. This enables us to apply (20.2) and obtain an immediate contradiction unless $n = 1$, 3 or 7. This completes the proof of (1.11).

21 · When is $P_{n,k}$ neutral?

This section is based on joint work with Sutherland [85]. Recall that d_n denotes the self-map of $P_{n,k}$ defined by reflection in the last coordinate hyperplane. We say that $P_{n,k}$ is neutral (elsewhere outsimple) if $d_n \simeq 1$, and define S-neutral similarly. If n and k are both odd then $P_{n,k}$ is neutral, as remarked in §7. If n is even then d_n has degree -1 on the integral homology $H_{n-1}(P_{n,k}) = Z$, and so $P_{n,k}$ is not S-neutral. Thus the interest resides in the case when n is odd and k even. Notice that $P_{k+1,k} = P^k$ is neutral for all even values of k. In the course of §6 we have already proved

Proposition (21.1). Suppose that $P_{n,k}$ is S-neutral, where n is odd and k even. Then $P_{m+n,k}$ and $P_{m+k-n,k}$ are S-neutral, whenever $m \equiv 0 \bmod \hat{a}_k$.

Here $\hat{a}_k$ is as in (1.10). Now consider $V_{n,k}$ as a Z_2-space under the outer automorphism which changes the sign of the last row and column of each matrix. Since the inclusion $P_{n,k} \to V_{n,k}$ is a Z_2-map it follows at once from (3.4) that $P_{n,k}$ is neutral if $V_{n,k}$ is neutral and $n \geq 2k$. However, the following result is more useful. Let n be odd and k even, so that $\lambda = \xi$, by (1.1). Let $m \equiv 0 \bmod \hat{a}_k$, with m large, so that there exists a homotopy-equivariant section $g : S^{m-1} \to P_{m,k}$. Let

$$\phi(1 \wedge g) : S(V_{n,k}) \wedge S^{m-1} \to P_{m+n,k}$$

be defined as in §7. Then $\phi(1 \wedge g)$ is homotopy-equivariant, by (7.6), and so the restriction

$$S(P_{n,k}) \wedge S^{m-1} \to P_{m+n,k}$$

is a homotopy-Z_2 equivalence. By composing $\phi(1 \wedge g)$ with a homotopy inverse of the restriction we obtain

Proposition (21. 2). Let n be odd and k even. Then $P_{n,k}$ is a homotopy-Z_2 S-retract of $V_{n,k}$. Hence $P_{n,k}$ is S-neutral if $V_{n,k}$ is neutral.

The relationship between neutrality and S-neutrality is clarified by

Proposition (21. 3). Let n be odd and k even. Then $P_{n,k}$ is neutral if and only if $P_{n,k}$ is S-neutral and $n > 2k$.

The 'if' part is suspension theory; there remains the problem of showing that $n > 2k$ whenever $P_{n,k}$ is neutral. After proving this, in the next few paragraphs, we go on to establish

Proposition (21. 4). Let n be odd and k even. If $P_{n,k}$ is S-neutral then either $n + 1$ or $k - n + 1$ is divisible by 2^t, where t is the least integer such that $2^t > k$.

Clearly (1. 12) follows at once from (21. 2) and (21. 4). As a preliminary to proving these results, consider the map $\psi : P_{n,1} \times S^1 \to P_{n+1,1}$ which is given by

$$\psi([x_1, \ldots, x_n], c_t) = [x_1, \ldots, x_{n-1}, x_n \cos \pi t, x_n \sin \pi t],$$

where $c_t = (\cos 2\pi t, \sin 2\pi t) \in S^1$. Since ψ maps $(P_{n,1} - e) \times (S^1 - e)$ homeomorphically onto $(P_{n+1,1} - e)$ it follows that

$$(21.5) \quad \psi^* : H^n(P_{n+1,1}) \approx H^n(P_{n,1} \times S^1)$$

in mod 2 cohomology. Next consider the homotopy $f : P_{n,k} \times I \to P_{n+1,k+1}$ of the inclusion $u : P_{n,k} \to P_{n+1,k+1}$ into ud_n which is given by

$$f([x_1, \ldots, x_n], t) = [x_1, \ldots, x_{n-1}, x_n \cos \pi t, x_n \sin \pi t].$$

Suppose that $d_n \simeq 1$ under a homotopy $f' : P_{n,k} \times I \to P_{n,k}$. We can fit f and uf' together, in the obvious way, to produce a homotopy

$$\theta : P_{n,k} \times S^1 \to P_{n+1,k+1},$$

such that $\theta(x, e) = ux$ $(x \in P_{n,k})$. By construction $q\theta \simeq \psi(q \times 1)$, as shown in the following diagram, where q denotes the natural projection.

$$\begin{array}{ccc} P_{n,k} \times S^1 & \xrightarrow{\theta} & P_{n+1,k+1} \\ {\scriptstyle q \times 1}\downarrow & & \downarrow{\scriptstyle q} \\ P_{n,1} \times S^1 & \xrightarrow[\psi]{} & P_{n+1,1} \end{array}$$

In mod 2 cohomology, therefore, we have

(21.6) $\theta^* : H^n(P_{n+1,k+1}) \approx H^n(P_{n,k} \times S^1)$.

Let α_r $(r = n - k, \ldots, n)$ generate $H^r(P_{n+1,k+1})$ and let β_s $(s = n - k, \ldots, n - 1)$ generate $H^s(P_{n,k})$. The conditions on θ imply that

$$(21.7)\quad \begin{cases} \text{(a)} & \theta^*\alpha_{n-k} = \beta_{n-k} \otimes 1, \\ \text{(b)} & \theta^*\alpha_n = \beta_{n-1} \otimes \gamma, \end{cases}$$

where γ is the generator of $H^1(S^1)$. If $n - k \leq r < n$ then $\theta^*\alpha_r$ equals either $\beta_r \otimes 1$ or $\beta_r \otimes 1 + \beta_{r-1} \otimes \gamma$. If further r is even then

$$Sq^1\theta^*\alpha_{r-1} = Sq^1(\beta_{r-1} \otimes 1),$$

by (21.7a) when $r = n - k + 1$ and since $Sq^1\beta_{r-2} = 0$ when $r > n-k+1$. Hence

(21.8) $\theta^*\alpha_r = \beta_r \otimes 1$ $(r$ even$)$.

If $n \leq 2k$ then $\alpha_n = \alpha_{n-k}.\alpha_k$, hence

$$\theta^*\alpha_n = (\beta_{n-k} \otimes 1) \,.\, (\beta_k \otimes 1) = 0,$$

contrary to (21.7b). This contradiction completes the proof of (21.3); we now begin the proof of (21.4).

For any space X we describe a pair x, y of mod 2 cohomology classes as <u>evenly connected</u> if $Sq^t x = y$, for some even integer t. This non-symmetric relation generates an equivalence relation; we describe

x, y as <u>evenly related</u> if they are equivalent in this sense. We prove

Lemma (21.9). <u>Let</u> n <u>be odd and</u> k <u>even. Then</u> α_{n-k} <u>and</u> α_n <u>are evenly related in</u> $H^*(P_{n+1,k+1})$ <u>unless either</u> n <u>or</u> $k - n$ <u>is congruent to</u> $-1 \bmod 2^t$, <u>where</u> t <u>is the least integer such that</u> $2^t > k$.

By (1.5) the following relations are valid, when defined:

$$Sq^4 \alpha_{8i-1} = \alpha_{8i+3}, \quad Sq^2 \alpha_{8i+3} = \alpha_{8i+5},$$
$$Sq^4 \alpha_{8i+5} = \alpha_{8i+9}, \quad \alpha_{8i+9} = Sq^2 \alpha_{8i+7}.$$

This shows that all the elements α_r with r odd and $n - k < r < n$ are evenly related. Let 2^s be the highest power of 2 dividing $n - k - 1$, so that $n - k \equiv 2^s + 1 \bmod 2^{s+1}$. Then

$$Sq^{2^s} \alpha_{n-k} = \alpha_{n-k+2^s},$$

and so α_{n-k} is evenly related to some other class of $P_{n+1,k+1}$ unless $2^s > k$. Let 2^t be the highest power of 2 dividing $n + 1$, so that $n \equiv 2^t - 1 \bmod 2^{t+1}$. Then

$$Sq^{2^t} \alpha_{n-2^t} = \alpha_n,$$

and so α_{n-k} is evenly related to some other class of $P_{n+1,k+1}$ unless $2^t > k$. Hence (21.9) follows at once.

We are now ready to prove (21.4), where n is odd and k even. Without real loss of generality we may suppose that $n > 2k$, since if necessary n can be increased by a multiple of $\hat{a}_k$. Therefore $P_{n,k}$ is neutral and so the map θ is defined. If α_n is evenly related to α_{n-k} then $\theta^* \alpha_n = 0$, by (21.7a) and naturality of the squaring operations. This contradicts (21.7b) and so establishes (21.4).

Finally, still following [85], we apply the results of §14 to prove

Theorem (21.10). <u>Suppose that the Whitehead square</u> $w_n \in \pi_{2n-1}(S^n)$ <u>can be halved, where</u> n <u>is odd. Then</u> $P_{n,k}$ <u>is neutral,</u> <u>for all even values of</u> k <u>such that</u> $n > 2k$.

To prove (21.10), consider the homotopy exact sequence

$$\pi_n(S^n) \xrightarrow{\Delta} \pi_{n-1}(V_{n,k}) \xrightarrow{u_*} \pi_{n-1}(V_{n+1,k+1}) \to 0.$$

Replacing (n, k) by $(n, n - k)$, in (14.5), we have $\Delta\iota_n = H_k w_n$, where

$$H_k : \pi_{2n-1}(S^n) \to \pi_{n-1}(V_{n,k}).$$

Since $w_n \in 2\pi_{2n-1}(S^n)$, by hypothesis, it follows that the kernel of u_* is contained in $2\pi_{n-1}(V_{n,k})$. Hence the kernel of

$$v_* : \pi_{n-1}(P_{n,k}) \to \pi_{n-1}(P_{n+1,k+1})$$

is contained in $2\pi_{n-1}(P_{n,k})$, by (3.2). Thus the corresponding coefficient homomorphism

$$v_\# : H^{n-1}(P_{n,k}, \pi_{n-1}(P_{n,k})) \to H^{n-1}(P_{n,k}, \pi_{n-1}(P_{n+1,k+1}))$$

is an isomorphism.

Now $v \simeq v'$ rel $P_{n-1,k-1}$, where $v' : P_{n,k} \to P_{n+1,k+1}$ is given by

$$v'[x_1, \ldots, x_{n-1}, x_n] = [x_1, \ldots, x_{n-1}, 0, x_n].$$

Consider the homotopy $h'_t : P_{n,k} \to P_{n+1,k+1}$ which is given by

$$h'_t[x_1, \ldots, x_{n-1}, x_n] = [x_1, \ldots, x_{n-1}\cos \pi t, x_{n-1}\sin \pi t, x_n].$$

By restricting h'_t to $P_{n-1,k-1}$ we obtain a homotopy $h''_t : P_{n-1,k-1} \to P_{n,k}$ such that $h''_0 = \iota$, the inclusion map, and $h''_1 = d\iota$, where d denotes reflection in the $(n - 1)$-st coordinate hyperplane. The obstruction to extending $h''_t | P_{n-2,k-2}$ to a homotopy of the identity into d is an element

$$\sigma \in H^{n-1}(P_{n,k}, \pi_{n-1}(P_{n,k})).$$

The obstruction to extending $vh''_t | P_{n-2,k-2}$ to a homotopy of v' into $v'd$ is the corresponding element

$$v'_\#\sigma \in H^{n-1}(P_{n,k}, \pi_{n-1}(P_{n+1,k+1})).$$

But $v'_{\#}\sigma = 0$, since $v'h''_t$ extends to the homotopy h'_t of v' into $v'd$, and so $\sigma = 0$, since $v'_{\#} = v_{\#}$, which is an isomorphism. Therefore $h''_t \mid P_{n-2, k-2}$ extends to a homotopy of the identity into d, and therefore $P_{n, k}$ is neutral, as asserted.

22 · When is $V_{n,2}$ neutral?

In this section and the next we shall prove the neutrality theorems for $V_{n,k}$, following [76] and [78]. First we take the case $k = 2$. We write $V_{n,2} = W^{2n-3}$, to emphasize dimension, and consider the map

$$h : S^{n-1} \times S^{n-1} \to SW^{2n-3}$$

defined as follows. If the geodesic distance between points $x, y \in S^{n-1}$ is equal to πt, where $0 < t < 1$, we define

$$h(x, y) = ((x, z), t),$$

where z is given by $y = x \cos \pi t + z \sin \pi t$. The diagonal points are mapped into the pole 0 of the suspension, and the antidiagonal into the pole 1. Since h is the projection of a trivialization of ΣW^{2n-3}, the fibre suspension, the degree of h is ± 1. Moreover h satisfies the condition

$$(22.1) \quad (Su) \circ h = v \circ h \circ (1 \times T),$$

as shown in the following diagram, where u is the self-map of W^{2n-3} which changes the sign of the second vector in each 2-frame, v is the self-map of SW^{2n-3} which reverses the suspension parameter, and T is the antipodal map on S^{n-1}.

$$\begin{array}{ccc} S^{n-1} \times S^{n-1} & \xrightarrow{1 \times T} & S^{n-1} \times S^{n-1} \\ h \downarrow & & \downarrow vh \\ SW^{2n-3} & \xrightarrow[Su]{} & SW^{2n-3} \end{array}$$

Now h is a map of type $(\pm\gamma, \pm\gamma)$, where γ generates $\pi_{n-1}(SW^{2n-3})$. By performing the Hopf construction on h we obtain an element θ, say, of

$\pi_{2n-1}(S^2W^{2n-3})$. If we use vh instead of h then, by a standard formula (see (2.22) of [64]) the element we obtain is equal to

$$-\theta + [S_*\gamma, S_*\gamma].$$

Also T has degree $(-1)^n$ and so (22.1) implies

$$(22.2) \quad (S^2u)_*\theta + (-1)^n\theta = \pm[S_*\gamma, S_*\gamma].$$

We use this relation to prove (1.13), but first some technical considerations are required.

Let $c : S^{r-1} \to S^{r-1}$ $(r = 2, 3, \ldots)$ be a map of degree 2 with mapping cone $Y^r = S^{r-1} \cup e^r$. Let n be odd. We regard W^{2n-3} as a complex of the form $Y^{n-1} \cup e^{2n-3}$ as described in §3. Recall that $S(S^{n-1} \times S^{n-1})$ is spherical, in dimension $2n - 1$, and so S^2W^{2n-3} is spherical, since Sh has degree ± 1. It follows at once that $Y^{n+1} = S^2Y^{n-1}$ is a retract of S^2W^{2n-3}. Now $\pi_{2n-1}(S^n) = S_*\pi_{2n-2}(S^{n-1})$, since n is odd, and hence $2\pi_{2n-1}(S^n)$ coincides with the kernel of the homomorphism

$$l_* : \pi_{2n-1}(S^n) \to \pi_{2n-1}(Y^{n+1})$$

induced by the inclusion. This proves that

$$(22.3) \quad \ker j_* = 2\pi_{2n-1}(S^n) \quad (n \text{ odd}),$$

as shown in the following diagram, where i, j are the inclusions.

$$\begin{array}{ccc} \pi_{2n-3}(S^{n-2}) & \xrightarrow{i_*} & \pi_{2n-3}(W^{2n-3}) \\ \downarrow S_*^2 & & \downarrow S_*^2 \\ \pi_{2n-1}(S^n) & \xrightarrow[j_*]{} & \pi_{2n-1}(S^2W^{2n-3}) \end{array}$$

Similarly $i_*c_*\phi = 0$, for any element $\phi \in \pi_{2n-3}(S^{n-2})$. However $c_*\phi = 2\phi + w_{n-2} \circ \phi'$, where $\phi' \in \pi_{2n-3}(S^{2n-5})$ denotes the generalized Hopf invariant of ϕ. Either $\phi' = 0$ or $\phi' = \eta_{2n-5} \circ \eta_{2n-4}$. Now $\Delta w_{n-1} = w_{n-2} \circ \eta_{2n-5}$, from (7.4) of [70], hence

$\Delta w_{n-1} \circ \eta_{2n-3} = w_{n-2} \circ \eta_{2n-5} \circ \eta_{2n-4}$, where $\Delta : \pi_{r+1}(S^{n-1}) \to \pi_r(S^{n-2})$ is the transgression. But $i_*\Delta = 0$, by exactness, and so $i_*(w_{n-2} \circ \phi') = 0$. Since $i_* c_* \phi = 0$ this shows that

$$(22.4) \quad 2 i_* \pi_{2n-3}(S^{n-2}) = 0.$$

We use this to prove

Proposition (22.5). Let $n \equiv 3 \bmod 4$ and $n \geq 11$. If $S^2_* i_* \phi = 0$, where $\phi \in \pi_{2n-3}(S^{n-2})$, then $i_*\phi = 0$.

We have $j_* S^2_* \phi = S^2_* i_* \phi = 0$, by hypothesis, and so $S^2_* \phi = 2\psi$, by (22.3), for some element $\psi \in \pi_{2n-1}(S^n)$. Since $n \equiv 3 \bmod 4$ we have $S_* \pi_{2n-3}(S^{n-2}) = \pi_{2n-2}(S^{n-1})$, hence $S^2_* \pi_{2n-3}(S^{n-2}) = S_* \pi_{2n-2}(S^{n-1}) = \pi_{2n-1}(S^n)$. Therefore $\psi = S^2_* \xi$, for some $\xi \in \pi_{2n-3}(S^{n-2})$, and so $\phi - 2\xi$ lies in the kernel of S^2_*. Hence $i_*\phi = [\sigma, \tau]$, by (14.5) and (22.4), for some $\sigma \in \pi_{n-2}(W^{2n-3})$ and $\tau \in \pi_n(W^{2n-3})$. Now $\pi_{n-2}(R_{n-2}) = Z_8$, as shown in [92], since $n \equiv 3 \bmod 4$ and $n \geq 11$. Since $\pi_n(W^{2n-3}) = Z_4$ this implies that $\Delta\tau \in 2\pi_{n-1}(R_{n-2})$. By (16.4) therefore $[\sigma, \tau] = -\langle \sigma, \Delta\tau \rangle = 0$, since $\sigma \in \pi_{n-2}(W^{2n-3}) = Z_2$, and so $i_*\phi = 0$, as asserted.

We are now ready to prove (1.13), which we restate as

Theorem (22.6). Let n be odd. Then W^{2n-3} is neutral if and only if the Whitehead square w_n can be halved.

If W^{2n-3} is neutral then $u \simeq 1$, hence $S^2 u \simeq 1$, hence $[S_*\gamma, S_*\gamma] = 0$, by (22.2). However $[S_*\gamma, S_*\gamma] = j_* w_n$, by naturality, and so $w_n \in 2\pi_{2n-1}(S^n)$, by (22.5).

Conversely, suppose that $w_n \in 2\pi_{2n-1}(S^n)$. If $n = 1$, 3 or 7 then W^{2n-3} is neutral, as we have seen. If $n \neq 1, 3, 7$ then $n \geq 11$ and $n \equiv 3 \bmod 4$, since $n + 1$ is a power of 2. Now $pu = p$, where $p : W^{2n-3} \to S^{n-1}$ is defined by taking the first vector of each 2-frame. Since n is odd there exists a deformation $h_t : Y^{n-1} \to W^{2n-3}$ of $u | Y^{n-1}$ into the inclusion such that ph_t is stationary. Since the complement of Y^{n-1} in W^{2n-3} is a $(2n - 3)$-cell the separation element

$$\delta = d(u, h_t, 1) \in \pi_{2n-3}(W^{2n-3})$$

is defined. By naturality

$$p_*\delta = d(pu, ph_t, p) = 0,$$

where $p_* : \pi_{2n-3}(W^{2n-3}) \to \pi_{2n-3}(S^{n-1})$, and so $\delta \in i_*\pi_{2n-3}(S^{n-2})$, by exactness. However

$$S_*^2\delta = d(S^2u, S^2h_t, 1) = (S^2u)_*\theta - \theta,$$

since $\theta \in \pi_{2n-1}(S^2W^{2n-3})$ is representable by maps of degree 1, and so

$$(22.7) \quad S_*^2\delta = \pm[S_*\gamma, S_*\gamma],$$

by (22.2). But $[S_*\gamma, S_*\gamma] = j_*w_n = 0$, by (22.3), since $w_n \in 2\pi_{2n-1}(S^n)$ by hypothesis. Therefore $S_*^2\delta = 0$, by (22.7). However $\delta \in i_*\pi_{2n-3}(S^{n-2})$, as we have seen, and so $\delta = 0$, from (22.5). Therefore h_t can be extended to a homotopy of u into the identity and so W^{2n-3} is neutral, as asserted.

23 · When is $V_{n,k}$ neutral?

The purpose of this section is to prove (1.14), which gives a necessary condition for the neutrality of $V_{n,k}$ when $k = 2$, 4 or 8. Some preliminary material about loop-spaces is required. We work in the category of pointed spaces and pointed maps but omit the word pointed throughout. As before we denote by $\pi(X, Y)$ the set of homotopy classes of maps $X \to Y$. We regard the suspension functor S and the loop functor Ω as adjoint, in the usual way. We regard $\pi(SX, Y)$ as a group, using track composition, and $\pi(X, \Omega Y)$ as a group, using loop composition, so that by taking adjoints we obtain a natural isomorphism

$$\xi : \pi(SX, Y) \approx \pi(X, \Omega Y).$$

We use the notation indicated in the following diagram for the structural maps associated with the product functor.

$$P \underset{p'}{\overset{p}{\leftrightarrows}} P \times Q \underset{q'}{\overset{q}{\rightleftarrows}} Q$$

Given elements $\alpha \in \pi(P, X)$, $\beta \in \pi(Q, X)$ we say that an element $\theta \in \pi(P \times Q, X)$ is of <u>type</u> (α, β) if $\alpha = p'^*\theta$, $\beta = q'^*\theta$. When P and Q are suspensions there is a well-known necessary and sufficient condition for the existence of such an element θ. Suppose that $P = SK$, $Q = SL$, where K, L are complexes. Consider the elements $\xi\alpha \in \pi(K, \Omega X)$, $\xi\beta \in \pi(L, \Omega X)$. Then (cf. (15.5)) we have

Proposition (23.1). <u>There exists an element</u> $\theta \in \pi(SK \times SL, X)$ <u>of type</u> (α, β) <u>if and only if</u> $p^*\xi\alpha$ <u>commutes with</u> $q^*\xi\beta$ <u>in the group</u> $\pi(K \times L, \Omega X)$.

Thus it is useful to determine sets of commuting elements in this group, as in

Proposition (23.2). Suppose that K is a suspension. Then the elements of the kernel of

$$q'^* : \pi(K \times L, \Omega X) \to \pi(L, \Omega X)$$

commute with one another.

For consider the complex M obtained from $K \times L$ by collapsing the axis $e \times L$ to a point. Since K is a suspension so also is M, hence $\pi(M, \Omega X)$ is abelian. However, the kernel of q'^* coincides with the image of

$$r^* : \pi(M, \Omega X) \to \pi(K \times L, \Omega X),$$

where r denotes the collapsing map, and so (23.2) follows at once.

Similarly, consider the self-map u of K which is given by track reversal (i.e. transforming the suspension parameter t into $1 - t$). Let v denote the self-map of the smash product $K \wedge L$ induced by the self-map $u \times 1$ of $K \times L$. Since v^* acts on $\pi(K \wedge L, \Omega X)$ by group inversion we obtain

Proposition (23.3). Suppose that K is a suspension. Then $(u \times 1)^*$ acts by group inversion on the intersection of the kernels of

$$\pi(K, \Omega X) \underset{p'_*}{\leftarrow} \pi(K \times L, \Omega X) \underset{q'_*}{\rightarrow} \pi(L, \Omega X).$$

Let Y be a space and let $j : Y \to \Omega SY$ denote the adjoint of the identity on SY. Write $v = \Omega u$, where $u : SY \to SY$ is given by track reversal. Then v_* constitutes an automorphism of the group $\pi(X, \Omega SY)$ for any space X. If $\theta \in \pi(X, Y)$ then

$$(23.4) \quad v_* j_*(\theta) = (j_*(\theta))^{-1},$$

from the definition of loop reversal, where

$$j_* : \pi(X, Y) \to \pi(X, \Omega SY).$$

Now let A, B, C be complexes and let h be a map as shown in the following diagram, where u denotes track reversal.

$$\begin{array}{ccc} SA \times SB & \xrightarrow{h} & SC \\ \downarrow{\scriptstyle u\times 1} & & \downarrow{\scriptstyle u} \\ SA \times SB & \xrightarrow[h]{} & SC \end{array}$$

With reference to the class $\gamma \in \pi(SA \times SB, SC)$ of h we prove

Proposition (23.5). Let $\gamma \in \pi(SA \times SB, SC)$ be an element of type (α, β), where $\alpha \in \pi(SA, SC)$, $\beta \in \pi(SB, SC)$. If $u_*(\gamma) = (u \times 1)^*\gamma$ then there exists an element $\hat{\gamma} \in \pi(S^2A \times S^2B, S^2C)$ of type $(S_*\alpha, S_*\beta)$.

Write $\gamma' = j_*\gamma$, where $j : SC \to \Omega S^2C$, and write $\alpha' = p^*j_*\alpha$, $\beta' = q^*j_*\beta$, so that $\alpha', \beta', \gamma' \in \pi(SA \times SB, \Omega S^2C)$. Then γ' is of type $(p'^*\alpha', q'^*\beta')$ and so, applying (23.2) twice over, we obtain

$$(23.6) \quad \alpha'(\beta'\gamma'^{-1}) = (\beta'\gamma'^{-1})\alpha' = \beta'(\gamma'^{-1}\alpha') = (\gamma'^{-1}\alpha')\beta'.$$

Consider the automorphisms $(u \times 1)^*$ and v_* of $\pi(SA \times SB, \Omega S^2C)$, where $v = \Omega u$. It follows at once from the hypothesis of (23.5) that these automorphisms agree on γ' and hence, by naturality, on α' and β'. Therefore

$$(23.7) \quad (u \times 1)^*(\alpha'\beta'\gamma'^{-1}) = v_*(\alpha'\beta'\gamma'^{-1}).$$

By (23.3) however

$$(u \times 1)^*(\alpha'\beta'\gamma'^{-1}) = \gamma'\beta'^{-1}\alpha'^{-1}.$$

Also it follows from (23.4) that v_* acts on α', β', γ' by group inversion, since these elements lie in the image of j_*, and so

$$v_*(\alpha'\beta'\gamma'^{-1}) = \alpha'^{-1}\beta'^{-1}\gamma'.$$

From these last three relations we obtain

$$\alpha'\beta'\gamma'^{-1} = \gamma'^{-1}\beta'\alpha' = \beta'\alpha'\gamma'^{-1},$$

by (23.6), and so α' commutes with β'. But

$$\alpha' = p^*\xi S_*\alpha, \quad \beta' = q^*\xi S_*\beta,$$

since $\xi S_* = j_*$, which commutes with p^*, q^*. Hence and from (23.1) we obtain (23.5).

From now on we work in terms of cohomology with mod 2 coefficients. Let $\gamma \in \pi(A \times B, X)$ be an element of type (α, β), where $\alpha \in \pi(A, X)$, $\beta \in \pi(B, X)$. We say that γ satisfies the H^r-condition $(r > 0)$ if both the induced homomorphisms

$$H^r(A) \xleftarrow{\alpha^*} H^r(X) \xrightarrow{\beta^*} H^r(B)$$

are non-trivial.

In particular, consider the real Stiefel manifold $V_{n,k}$ as an $(n - k)$-sphere bundle over $V_{n,k-1}$, in the usual way. Recall from §12 that the $(k - 1)$-fold fibre suspension $\Sigma^{k-1}V_{n,k}$ is trivial, as an $(n - 1)$-sphere bundle. Choose the trivialization

$$f : \Sigma^{k-1}V_{n,k} \to S^{n-1} \times V_{n,k-1}$$

given by the retraction ρ as in (12.4), (12.5) and let

$$h : S^{n-1} \times V_{n,k-1} \to S^{k-1}V_{n,k}$$

be defined by composing f^{-1} with the natural projection of the fibre suspension onto the ordinary suspension. I assert that the homotopy class γ of h satisfies the H^{n-1}-condition.

Since ρ is a retraction it follows that h maps the axis $S^{n-1} \times e$ homeomorphically onto $S^{k-1}V_{n-k+1,1} \subset S^{k-1}V_{n,k}$, and hence that α generates $\pi_{n-1}(S^{k-1}V_{n,k})$. Thus α^* is non-trivial. Suppose, to obtain a contradiction, that β^* is trivial. Then $\operatorname{im} p^* \subset \operatorname{im} h^*$, where

$$H^{n-1}(S^{k-1}V_{n,k}) \xrightarrow{h^*} H^{n-1}(S^{n-1} \times V_{n,k-1}) \xleftarrow{p^*} H^{n-1}(S^{n-1}).$$

But hg is constant, where $g = f|S^{k-2} \times V_{n,k-1}$, and so $g^*h^* = 0$. Hence $g^*p^* = 0$, where

$$H^{n-1}(S^{n-1}) \xrightarrow{p^*} H^{n-1}(S^{n-1} \times V_{n,k-1}) \xrightarrow{g^*} H^{n-1}(S^{k-2} \times V_{n,k-1}).$$

On the other hand $pg = \rho | S^{k-2} \times V_{n,k-1}$ and so pg, by (12.5), maps the axis $e \times V_{n,k-1}$ onto S^{n-1} as in the standard fibration. Hence $g^*p^* \neq 0$ and we have a contradiction. Therefore β^* is non-trivial and so γ, the homotopy class of h, satisfies the H^{n-1}-condition, as asserted.

These results are now used to prove

Proposition (23.8). Let n be odd, k even and $n \geq 2k - 2$. Suppose that $V_{n,k}$ is neutral. Then there exists an element

$$\hat{\theta} \in \pi(S^n \times SP_{n-1,k-1}, S^k V_{n,k})$$

which satisfies the H^n-condition.

Since $P_{n,k-1}$ is $(n - 1)$-dimensional and $(n - k)$-connected the condition $n \geq 2k - 2$ ensures that $P_{n,k-1}$ can be desuspended. Hence (23.5) applies to the class

$$\theta \in \pi(S^{n-1} \times P_{n,k-1}, S^{k-1} V_{n,k})$$

of the restriction of h, and shows that there exists an element

$$\hat{\theta} \in \pi(S^n \times SP_{n,k-1}, S^k V_{n,k})$$

of type $(S_*\alpha, S_*\beta_0)$, where (α, β_0) is the type of θ. Since h satisfies the H^{n-1}-condition so does θ, by (3.2). Hence $\hat{\theta}$ satisfies the H^n-condition, which proves (23.8).

For any cellular map $f : S^{n+2} P_{n,k-1} \to SP_{n+k,k}$ the mapping cone $K = C_f$ is a $(2n + 2)$-dimensional n-connected complex. The mod 2 cohomology of K is given by $H^r(K) = Z_2$ for $n + 1 \leq r \leq n + k$ and $2n - k + 4 \leq r \leq 2n + 2$ while $H^r(K) = 0$ for $n + k < r < 2n - k + 4$. Of course the action of the Steenrod squares depends on f, as well as on n and k. We prove

Proposition (23.9). Let $k = 2$, 4 or 8. Let n be odd and let $n \geq 2k - 2$. If $V_{n,k}$ is neutral then there exists a cellular map

$$f : S^{n+2} P_{n,k-1} \to SP_{n+k,k}$$

of which the mapping cone $K = C_f$ has the property that

$$Sq^{n+1} : H^{n+1}(K) \approx H^{2n+2}(K).$$

To prove (23.9) we first observe that there exists a map $f : S^k V_{n,k} \to V_{n+k,k}$ such that

$$f^* : H^n(V_{n+k,k}) \approx H^n(S^k V_{n,k}).$$

In fact f can be taken to be the composition

$$S^{k-1} * V_{n,k} \xrightarrow{t*1} V_{k,k} * V_{n,k} \xrightarrow{h} V_{n+k,k},$$

where t is a cross-section and h the intrinsic map. From (23.8) the element

$$f^*\hat{\theta} \in \pi(S^n \times SP_{n,k-1}, V_{n+k,k})$$

satisfies the H^n-condition. Since the pair $(V_{n+k,k}, P_{n+k,k})$ is 2n-connected and since $S^n \times SP_{n-1,k-1}$ has dimension $2n$ it follows that $f_*\hat{\theta}$ lies in the image of

$$i_* : \pi(S^n \times SP_{n,k-1}, P_{n+k,k}) \to \pi(S^n \times SP_{n,k-1}, V_{n+k,k}),$$

where i denotes the inclusion. Moreover ϕ satisfies the H^n-condition, where $\phi = i_*^{-1} f_* \hat{\theta}$. By performing the Hopf construction on ϕ we obtain an element $\psi \in \pi(S^{n+2}P_{n,k-1}, SP_{n+k,k})$. Consider the mapping cone K of a representative f of ψ. Since $H^{2n}(P_{n+k,k}) = 0$ and since ϕ satisfies the H^n-condition it follows easily (see (1.4) of [141]) that Sq^{n+1} is an isomorphism, as asserted.

Finally we use this result and the Adem relations to prove (1.14), which it is convenient to restate as

Theorem (23.10). Let n be odd and let $n \geq 2k - 2$ where $k = 2$, 4 or 8. If $V_{n,k}$ is neutral then $n + 1$ is a power of 2.

When $k = 2$ the conclusion is immediate. When $k = 4$ the Adem theorem shows that either $n + 1$, $n - 1$ or $n - 3$ is a power of 2. But $n \equiv 5$ or 7 mod 8, by (1.12), and so we have a contradiction unless $n + 1$ is a power of 2. When $k = 8$ the Adem theorem shows that one of $n + 1$ $n - 1, \dots, n - 11$ is a power of 2. Also $n \equiv 7$ or 15 mod 16, by (1.12).

If $n \equiv 7 \bmod 16$ and $n \geq 23$ a brief calculation, using (3.1), shows that the Steenrod algebra contains no element of degree $n + 1$ which acts non-trivially. Hence $n + 1$ is a power of 2, as asserted.

24 · Further results and problems

In this final section I give a brief account of a few of the many other topics I should like to have covered in this volume. The small collection of problems at the end may also serve to introduce some further material; a good many unsolved problems have already been mentioned in the text.

I have already referred, in §2 above, to the literature on the cohomology theory of Stiefel manifolds, especially Chapter IV of the Steenrod-Epstein memoir [134]. The corresponding results for complex K-theory were obtained by Gitler and Lam [50], but unfortunately these have yet to find an adequate application. It may be that additional computation of the Adams operations, or the extension of their work to real K-theory, might be fruitful, particularly in relation to the neutrality and triviality problems. The Mahowald memoir [100] is the main source of information about the homotopy groups in the real case; some other useful references are given in the bibliography.

The Lie groups Sp_n and R_{2n+1} $(n \geq 1)$ have isomorphic cohomology rings over Z_p (p odd), moreover the isomorphism is compatible with the reduced power operations. This observation led Serre to conjecture that Sp_n and R_{2n+1} have isomorphic homotopy groups mod $\mathcal{C}$, where $\mathcal{C}$ denotes the class of 2-primary groups. When $n = 1$ this is clear, since of course Sp_1 is the universal covering group of R_3. Serre's conjecture was first proved by Bruno Harris [54], who obtained a number of other results of a similar nature. Still working mod $\mathcal{C}$ he showed that the following two exact sequences are short exact and split:

$$\pi_r(V_{2n+1,2k}) \to \pi_r(W_{2n+1,2k}) \to \pi_r(W_{2n+1,2k}, V_{2n+1,2k}),$$

$$\pi_r(X_{n,k}) \to \pi_r(W_{2n,2k}) \to \pi_r(W_{2n,2k}, X_{n,k}).$$

Eventually this leads to the conclusion that

$$\pi_r(X_{n,k}) \approx \pi_r(V_{2n+1,2k}) \pmod{\mathcal{C}},$$

a generalization of Serre's conjecture.

Recently these results have been re-examined by Friedlander [45] using localization theory. Away from the prime 2 the Harris argument shows that Sp_n and R_{2n+1} have the same homotopy type, also that $\Omega X_{n,k}$ and $\Omega V_{2n+1,2k}$ have the same homotopy type. Using methods inspired by algebraic geometry Friedlander has shown that $X_{n,k}$ and $V_{2n+1,2k}$ have the same homotopy type, away from the prime 2. The Harris results also suggest other possibilities. For example, is $X_{n,k}$ a retract of $W_{2n,2k}$, and is $V_{2n+1,2k}$ a retract of $W_{2n+1,2k}$, in the same localized sense? It would also be interesting to look at the triviality problem for complex and quaternionic Stiefel manifolds from the local point of view.

A smooth manifold M is said to be parallelizable if the tangent bundle $T(M)$ is trivial. Sutherland [139] was the first to show that $O_{n,k}$ is parallelizable for $k \geq 2$. Other proofs of this have been given by Handel [53] and Lam [94] provided $k > 2$ in the real case. As we have seen in §12, it is easy to establish parallelizability in the stable sense.

If the manifold M is an H-space then $T(M)$ is J-trivial, as shown by Browder and Spanier [33]. For $k \geq 2$ it seems plausible to conjecture that neither $V_{n,k}$ nor $W_{n,k}$ is an H-space when $k < n - 1$, also that $X_{n,k}$ is not an H-space when $k < n$. The case of $X_{n,n-1} = Sp_n/Sp_1$ appears to present difficulty, but in all the other cases the conjecture can be established fairly easily by cohomological methods (I am most grateful to Dr. Hubbuck for information on this point).

The homotopy groups of Stiefel manifolds play a basic role in the index theory of singular vector fields. To be precise, if M is oriented and $\dim M = n$ then the index of a k-field X on M, with finitely many singularities, is an element $I(X) \in \pi_{n-1}(V_{n,k})$. Reversing the orientation of M changes $I(X)$ into $-\lambda_* I(X)$, where λ is as in §1 etc. Thomas [142] has surveyed the literature in this area to which Atiyah and Dupont [10] and Dupont [35] have recently made major contributions. Incidentally both [10] and [87] contain some information about $\pi_{n-1}(V_{n,k})$ when $n \equiv 0 \bmod a_k$.

Stiefel manifolds are fundamental to the theory of immersions [57], submersions, etc. These applications underline the importance of studying, for any pair U, V of euclidean bundles over a space X, the fibre bundle formed from morphisms $U_x \to V_x$ $(x \in X)$, and particularly the subbundles formed by the monomorphisms and epimorphisms. Many of the topics we have been studying can be generalized to these 'Stiefel bundles'; further details, with references to the literature, can be found in the Oxford D. Phil. theses of M. C. Crabb and L. Woodward.

Problems (mainly taken from the literature)

1. Show that the manifold $O^*_{n,k}$ of linearly independent k-frames is homeomorphic to $O_{n,k} \times A^m$, where $m = \frac{1}{2}k(k-1)$.

2. Show that $O_{m,k}$ is contractible in $O_{m+n,k}$ if and only if $m \leq n$.

3. Suppose that S^{n-1} admits a $(k-1)$-field, where $2k \leq n+1$. Show (see [62]) that every field of tangent $(k-1)$-planes can be spanned by a field of tangent $(k-1)$-frames if and only if $\pi_{n-1}(O_k) = 0$.

4. The projective Stiefel manifold $V_{n,k}/Z_2$ is defined by identifying each k-frame $(v_1, \ldots, v_k)$ with $(-v_1, \ldots, -v_k)$. Investigate the cohomology of this space (see [49]).

5. Show that $P_{n,k}$ is S-neutral if and only if there exists an S-map $SP_{n,k} \to P_{n+1,k+1}$ which induces an isomorphism of mod 2 cohomology in dimension n.

6. In the function-space of maps $f : S^k \to S^n$, where $1 \leq k < n$, let $\mathcal{S}(n, k)$ denote the subspace of maps f such that

$$\|fx - fy\| \leq K\|x - y\| \qquad (x, y \in S^k)$$

for some $K \in I$ depending on f. Show (see [102]) that $\mathcal{S}(n, k)$ is homeomorphic to the double mapping cylinder of the fibrations

$$V_{n+1,k+1} \leftarrow V_{n+1,k+2} \to S^n.$$

7. Show (see [70]) that the Hurewicz homomorphism

$$\pi_{2n-3}(V_{n,2}) \to H_{2n-3}(V_{n,2}) \quad (n \text{ odd}, \ n \geq 5)$$

has index 4 or 8 according as $n \equiv 1$ or $3 \bmod 4$.

8. Calculate the index of the image of

$$i_* : \pi_{2n-3}(V_{n,2}) \to \pi_{2n-3}(W_{n,2}),$$

where n is odd and i denotes inclusion.

9. Consider the standard involution T of $W_{2n,2k}$ which has $X_{n,k}$ as fixed-point set. Show that if $\theta \in \pi_r(W_{2n,2k}, X_{n,k})$, for any r, then

$$2^{2k-1}(\theta + T_*\theta) = 0.$$

[Hint: the case $k = 1$ follows from a result proved in [65].]

Deduce that if $\phi \in \pi_r(X_{n,k})$ is an element such that $l_*\phi = 0$ in $\pi_r(W_{2n,2k})$, where l denotes the inclusion, then $4^k\phi = 0$.

10. Let $1 < l < k \leq n$. Show (see [138]) that $W_{n,k}$ does not have a cross-section over $W_{n,l}$ unless $k = n$ and $l = n - 1$. Also show that $X_{n,k}$ does not have a cross-section over $X_{n,l}$.

11. The number $U(n,k)$ is defined to be the index of $p_*\pi_{2n-1}(W_{n,k})$ in $\pi_{2n-1}(W_{n,1})$. Show that

(i) $U(n, k)$ is a multiple of $U(n, l)$, for $l \leq k$;

(ii) $U(n, k).U(m, k)$ is a multiple of $U(m + n, k)$;

(iii) If $U(n, k) = 1$ and $m \geq 2k - 1$ then $U(m, k) = U(m+n, k)$.

12. With the notation of Problem 11, show that $U(n, 3)$ is given by the following table (see [128])

$n \equiv$ (mod 24)	3	4	5	6	7	8	9	10	11	12	13	14
$U(n, 3)$	2	6	24	4	12	3	8	12	6	2	24	12
$n \equiv$ (mod 24)	15	16	17	18	19	20	21	22	23	24	25	26
$U(n, 3)$	4	3	24	4	6	6	8	12	12	1	24	12

13. The action of O_n on $V_{n,k}$ is given by a map

$$g : O_n \times V_{n,k} \to V_{n,k}.$$

The principal fibration $O_{n+1} \to S^n$ is classified by a map $f : S^{n-1} \to O_n$. Show that the fibration $V_{n+1,k+1} \to S^n$ is trivial, in the sense of fibre homotopy type, if and only if

$$g(f \times 1) \simeq \rho : S^{n-1} \times V_{n,k} \to V_{n,k},$$

where ρ denotes the right projection.

14. Show that if $1 < k < n$ then neither of the fibrations

$$W_{n,k} \to W_{n,1}, \quad X_{n,k} \to X_{n,1}$$

is trivial, in the sense of fibre homotopy type.

15. Show (see [78]) that the dk-fold suspension of the fundamental class of $O_{n,k}$ is a spherical class.

Bibliography

1. J. F. Adams. On the non-existence of elements of Hopf invariant one, Ann. of Math., 72 (1960), 20-104.
2. J. F. Adams. Vector fields on spheres, Topology, 1 (1962), 63-5.
3. J. F. Adams. Vector fields on spheres, Ann. of Math., 75 (1962), 603-32.
4. J. F. Adams. On the groups J(X) I-IV, Topology, 2 (1963), 181-95; 3 (1965), 137-72, 193-222; 5 (1966), 21-71.
5. J. F. Adams and G. Walker. On complex Stiefel manifolds, Proc. Cambridge Phil. Soc., 61 (1965), 81-103.
6. J. Adem. The iteration of the Steenrod squares in algebraic topology, Proc. Nat. Acad. Sci., 38 (1952), 720-6.
7. M. F. Atiyah. Thom complexes, Proc. London Math. Soc., 11 (1961), 291-310.
8. M. F. Atiyah. K-theory, Benjamin, New York, 1967.
9. M. F. Atiyah, R. Bott and A. Shapiro. Clifford modules, Topology, 3 (Suppl. 1) (1964), 3-38.
10. M. F. Atiyah and S. L. Dupont. Vector fields with finite singularities, Acta Math., 128 (1972), 1-40.
11. M. F. Atiyah and J. A. Todd. On complex Stiefel manifolds, Proc. Cambridge Phil. Soc., 56 (1960), 342-353.
12. W. D. Barcus. Note on cross-sections over CW-complexes, Quart. J. Math. (Oxford), 5 (1954), 150-60.
13. W. D. Barcus and M. G. Barratt. On the homotopy classification of the extensions of a fixed map, Trans. Amer. Math. Soc., 88 (1958), 57-74.
14. M. G. Barratt. Homotopy ringoids and homotopy groups, Quart. J. of Math. (Oxford), 5 (1954), 271-90.

15. M. G. Barratt. Note on a formula due to Toda, J. London Math. Soc., 36 (1961), 95-6.

16. M. G. Barratt and P. J. Hilton. On join operations in homotopy groups, Proc. London Math. Soc., 3 (1953), 430-45.

17. M. G. Barratt and M. E. Mahowald. The metastable homotopy of O(n), Bull. Amer. Math. Soc., 70 (1964), 758-60.

18. M. G. Barratt and G. H. Paechter. A note on $\pi_r(V_{n,m})$, Proc. Nat. Acad. Sci., 38 (1952), 119-21.

19. P. Baum. On the cohomology of homogeneous spaces, Topology, 7 (1968), 15-38.

20. P. Baum and W. Browder. The cohomology of quotients of classical groups, Topology, 3 (1965), 305-36.

21. J. C. Becker and D. H. Gottlieb. The transfer map and fibre bundles, Topology, 14 (1975), 1-12.

22. A. L. Blakers and W. S. Massey. Products in homotopy theory, Ann. of Math., 58 (1953), 295-328.

23. A. Borel. Sur la cohomologie des variétés de Stiefel et de certains groupes de Lie, C. R. Acad. Sci. (Paris), 208 (1939), 1263-5.

24. A. Borel. Sur la cohomologie des espaces fibrés principaux et des espaces homogènes de groupes de Lie compacts, Ann. of Math., 57 (1953), 115-207.

25. A. Borel. Sur la cohomologie des variétés de Stiefel, Comment. Math. Helv., 37 (1963), 239-40.

26. A. Borel and J.-P. Serre. Groupes de Lie et puissances reduites de Steenrod, Amer. J. of Math., 75 (1953), 409-48.

27. R. Bott. The space of loops on a Lie group, Michigan Math. J., 5 (1958), 35-61.

28. R. Bott. The stable homotopy of the classical Lie groups, Ann. of Math., 70 (1959), 313-37.

29. R. Bott. Quelques remarques sur les theorèmes de periodicité, Bull. Soc. Math. France, 87 (1959), 293-310.

30. R. Bott. A note on the Samelson product in the classical groups, Comment. Math. Helv., 34 (1960), 249-56.

31. R. Bott. Lectures on K(X), Harvard, 1962.

32. R. Bott. A note on the KO-theory of sphere-bundles, Bull. Amer. Math. Soc., 68 (1962), 395-400.

33. W. Browder and E. Spanier. H-spaces and duality, Pacific J. of Math., 12 (1962), 411-14.

34. A. Dold. Partitions of unity in the theory of fibrations, Ann. of Math., 78 (1963), 223-55.

35. J. L. Dupont. K-theory obstructions to the existence of vector fields, Acta Math., 133 (1974), 67-80.

36. B. Eckmann. Zur homotopietheorie gefaserter Räume, Comment. Math. Helv., 14 (1941/2), 141-92.

37. B. Eckmann. Gruppentheoretischer Beweis des Satzes von Hurwitz-Radon uber der Komposition quadratischer Formen, Comment. Math. Helv., 15 (1942), 358-36.

38. B. Eckmann. Stetige Lösungen linearer Gleichungsysteme, Comment. Math. Helv., 15 (1942/3), 318-39.

39. C. Ehresmann. Sur la topologie de certains espaces homogènes, Ann. of Math., 35 (1934), 396-443.

40. C. Ehresmann. Sur la topologie de certaines variétés algebriques réelles, J. Math. Pures et Appl., 16 (1937), 69-100.

41. C. Ehresmann. Sur la variété des génératrices planes d'une quadrique réelle et sur la topologie du groupe orthogonal à n variables, C. R. Acad. Sci. (Paris), 208 (1939), 321-3.

42. S. Feder and S. Gitler. Stable homotopy types of stunted complex projective spaces, Proc. Cambridge Phil. Soc., 73 (1973), 431-8.

43. S. Feder and S. Gitler. Stable homotopy types of Thom complexes, Quart. J. of Math. (Oxford), 25 (1974), 143-9.

44. H. Federer. A study of function spaces by spectral sequences, Trans. Amer. Math. Soc., 82 (1956), 340-61.

45. E. M. Friedlander. Maps between localized homogeneous spaces, Topology (to appear).

46. M. Fujii. KO-groups of complex projective spaces, Osaka J. Math., 4 (1967), 141-9.

47. M. Gilmore. Some Whitehead products on odd spheres, Proc. Amer. Math. Soc., 20 (1969), 375-7.

48. S. Gitler. The projective Stiefel manifolds II, Applications. Topology, 7 (1968), 47-53.

49. S. Gitler and D. Handel. The projective Stiefel manifolds I, Topology, 7 (1968), 39-46.

50. S. Gitler and K. Y. Lam. The K-theory of Stiefel manifolds, The Steenrod algebra and its applications, Lecture notes in mathematics, 168 (Springer 1970).

51. A. Haefliger. Differentiable embeddings of S^n in S^{n+q} for $q > 2$, Ann. of Math., 83 (1966), 402-36.

52. A. Haefliger and M. W. Hirsch. Immersions in the stable range, Ann. of Math., 75 (1962), 231-41.

53. D. Handel. Note on the parallelizability of real Stiefel manifolds, Proc. Amer. Math. Soc., 16 (1965), 1012-14.

54. B. Harris. Suspensions and characteristic maps for symmetric spaces, Ann. of Math., 76 (1962), 295-305.

55. B. Harris. Some calculations of homotopy groups of symmetric spaces, Trans. Amer. Math. Soc., 106 (1963), 174-84.

56. P. J. Hilton. On the homotopy groups of the union of spheres, J. London Math. Soc., 30 (1955), 154-72.

57. M. W. Hirsch. Immersions of manifolds, Trans. Amer. Math. Soc., 93 (1959), 242-76.

58. C. S. Hoo and M. E. Mahowald. Some homotopy groups of Stiefel manifolds, Bull. Amer. Math. Soc., 71 (1965), 661-7.

59. A. Hurwitz. Über die Komposition quadratischer Formen von beliebig vielen Variabeln, Nachr. Ges. d. Wiss. Göttingen (1898), 309-16.

60. A. Hurwitz. Über die Komposition der quadratischer Formen, Math. Ann., 88 (1923), 1-25.

61. S. Y. Husseini. A note on the intrinsic join of Stiefel manifolds, Comment. Math. Helv., 38 (1963), 26-30.

62. I. M. James. Note on factor spaces, J. London Math. Soc., 28 (1953), 278-85.

63. I. M. James. On the iterated suspension, Quart. J. Math. (Oxford), 5 (1954), 1-10.

64. I. M. James. On the suspension triad, Ann. of Math., 63 (1956), 191-247.

65. I. M. James. On the suspension sequence, Ann. of Math., 65 (1957), 74-107.

66. I. M. James. Whitehead products and vector fields on spheres, Proc. Cambridge Phil. Soc., 53 (1957), 817-20.

67. I. M. James. The intrinsic join, Proc. London Math. Soc., 8 (1958), 507-35.

68. I. M. James. Cross-sections of Stiefel manifolds, Proc. London Math. Soc., 8 (1958), 536-47.

69. I. M. James. Spaces associated with Stiefel manifolds, Proc. London Math. Soc., 9 (1959), 115-40.

70. I. M. James. Products on spheres, Mathematika, 6 (1959), 1-13.

71. I. M. James. On H-spaces and their homotopy groups, Quart. J. Math. (Oxford), 11 (1960), 161-79.

72. I. M. James. Suspension of transgression, Fundamenta Math., 50 (1962), 501-7.

73. I. M. James. The space of bundle maps, Topology, 2 (1963), 45-59.

74. I. M. James. On the homotopy theory of the classical groups, An. da Acad. Brasileira de Ciências (1967), 39-44.

75. I. M. James. On the Bott suspension, J. of Math. Kyoto Univ., 9 (1969), 161-88.

76. I. M. James. Note on Stiefel manifolds I, Bull. London Math. Soc., 2 (1970), 199-203.

77. I. M. James. On the homotopy-symmetry of sphere-bundles, Proc. Cambridge Phil. Soc., 69 (1971), 291-4.

78. I. M. James. Note on Stiefel manifolds II, J. London Math. Soc., 4 (1971), 109-17.

79. I. M. James. On the homotopy type of Stiefel manifolds, Proc. Amer. Math. Soc., 29 (1971), 151-8.

80. I. M. James. Products in homotopy groups, Compositio Math., 23 (1971), 329-45.

81. I. M. James. Which fibre spaces are decomposable? Indag. Math., 37 (1975), 385-90.

82. I. M. James. Homotopy-equivariance, Comment. Math. Helv., 50 (1975), 521-33.

83. I. M. James. Relative Stiefel manifolds, J. London Math. Soc., 13 (1976), 331-5.

84. I. M. James. On complex Stiefel manifolds, (to appear).

85. I. M. James and W. A. Sutherland. On stunted real projective spaces, Quart. J. of Math. (Oxford), 25 (1974), 101-12.

86. I. M. James and E. Thomas. Homotopy-commutativity in rotation groups, Topology, 1 (1962), 121-4.

87. I. M. James, E. Thomas, H. Toda and G. W. Whitehead. On the symmetric square of a sphere, J. of Math. and Mech., 5 (1963), 771-6.

88. I. M. James and J. H. C. Whitehead. Note on fibre spaces, Proc. London Math. Soc., 4 (1954), 129-37.

89. I. M. James and J. H. C. Whitehead. On the homotopy theory of sphere-bundles over spheres, Proc. London Math. Soc., 4 (1954), 196-218; 5 (1955), 148-66.

90. M. Karoubi. Fondements de la K-theorie, Fac. Sci. d'Alger. (1966/7).

91. M. Kervaire. Non-parallelizability of the n-sphere for $n > 7$, Proc. Nat. Acad. Sci., 44 (1958), 280-3.

92. M. Kervaire. Some nonstable homotopy groups of Lie groups, Ill. J. of Math., 4 (1960), 161-9.

93. A. Kirchhoff. Sur l'existence de certains champs tensoriels sur les sphères à n dimensions, C. R. Acad. Sci. (Paris), 223 (1948), 1258-60.

94. K. Y. Lam. A formula for the tangent bundle of flag manifolds and related manifolds, Trans. Amer. Math. Soc., 213 (1975), 305-11.

95. J. Leise. The Bott suspension and intrinsic join, (preprint).

96. G. S. McCarty, Jr. Products between homotopy groups and the J-morphism, Quart. J. of Math. (Oxford), 15 (1964), 362-70.

97. G. S. McCarty, Jr. The value of J at a Samelson product, Proc. Amer. Math. Soc., 19 (1968), 164-7.

98. M. E. Mahowald. A short proof of the James periodicity of $\pi_{k+p}(V_{k+m,m})$, Proc. Amer. Math. Soc., 16 (1965), 512.

99. M. E. Mahowald. A Samelson product in SO(2n), Bol. Soc. Mat. Mexicana, (1966), 80-3.

100. M. E. Mahowald. The metastable homotopy of O(n), Amer. Math. Soc. Memoir, 72 (1967).

101. M. E. Mahowald and R. J. Milgram. Operations which detect Sq^4 in connective K-theory and their applications, (to appear).

102. H. J. Marcum. Stiefel manifolds and similarity mappings of spheres, Atas do 10° Coloquio Brasileiro de Matématica.

103. H. Matsunaga. On the homotopy groups of Stiefel manifolds, Mem. Fac. Sci. Kyushu Univ., 13 (1959), 152-6.

104. H. Matsunaga. On the groups $\pi_{2n+i}U(n)$ for $i = 3, 4, 5$, Mem. Fac. Sci. Kyushu Univ., 15 (1961), 72-81.

105. H. Matsunaga. Corrections of the preceding paper and note on the James number, Mem. Fac. Sci. Kyushu Univ., 16 (1962), 60-1.

106. H. Matsunaga. On the groups $\pi_{2n+7}U(n)$ odd primary components, Mem. Fac. Sci. Kyushu Univ., 16 (1962), 66-74.

107. H. Matsunaga. Some stunted projective spaces, Mem. Fac. Sci. Kyushu Univ., 16 (1962), 75-87.

108. H. Matsunaga. Applications of functional fohomology operations to the calculus of $\pi_{2n+i}U(n)$ for $i = 6$ and 7, Mem. Fac. Sci. Kyushu Univ., 17 (1963), 29-62.

109. J. P. May. On kO-oriented bundle theories, (preprint).

110. R. J. Milgram and P. Zvengrowski. Projective Stiefel manifolds and skew-linear vector fields, Proc. London Math. Soc., 28 (1974), 671-82.

111. R. J. Milgram and P. Zvengrowski. Skewness of r-fields on spheres, Topology, (to appear).

112. C. E. Miller. The topology of rotation groups, Ann. of Math., 57 (1953), 90-113.

113. J. Milnor and E. Spanier. Two remarks on fiber homotopy type, Pacific J. of Math., 10 (1960), 585-90.

114. M. Mori. Homotopy groups of Stiefel manifolds, Mem. Fac. Sci. Kyushu Univ., 25 (1971), 304-17.

115. Y. Nomura. Exact squares and some theorems of Toda, Science reports Coll. Educ. Tokyo Univ., 23 (1974), 1-12.

116. G. F. Paechter. The groups $\pi_r(V_{n,m})$ I-V, Quart. J. Math. (Oxford), 7 (1956), 249-68; 9 (1958), 8-27; 10 (1959), 17-37, 241-60; 11 (1960), 1-16.

117. D. Quillen. The Adams conjecture, Topology, 10 (1971), 67-80.

118. J. Radon. Linear Scharen orthogonaler Matrizen, Abh. Sem. Hamburg, 1 (1923), 1-14.

119. A. D. Randall. Cohomology of quasi-projective Stiefel manifolds, Rocky Mountain J. of Math., 3 (1973), 619-30.

120. Y. Saito. On the homotopy groups of Stiefel manifolds, J. Inst. Poly. Osaka City Univ., 6 (1955), 39-45.

121. H. Samelson. A connection between the Whitehead and Pontrjagin product, Amer. J. of Math., 75 (1953), 744-52.

122. H. Samelson. Groups and spaces of loops, Comment. Math. Helv., 28 (1954), 278-86.

123. B. Sanderson. Immersions and embeddings of projective spaces, Proc. London Math. Soc., 14 (1964), 137-53.

124. H. Scheerer. Transitive actions on Hopf homogeneous spaces, Manuscripta Math., 4 (1971), 99-134.

125. J.-P. Serre. Homologie singulière des espaces fibrés, Ann. of Math., 54 (1951), 425-505.

126. J.-P. Serre. Groupes d'homotopie et classes de groupes abeliens, Ann. of Math., 58 (1953), 258-94.

127. G. B. Segal. Equivariant K-theory, Publ. Math. IHES (Paris), 34 (1968), 129-51.

128. F. Sigrist. Groupes d'homotopie des variétés de Stiefel complexes, Comment. Math. Helv., 43 (1968), 121-31.

129. F. Sigrist. Determination des groupes d'homotopie $\pi_{2k+7}(U_{k+m,m})$, C. R. Acad. Sci. (Paris), 269 (1969), 1061-2.

130. F. Sigrist and U. Suter. Cross-sections of symplectic Stiefel manifolds, Trans. Amer. Math. Soc., 184 (1973), 247-59.

131. E. Spanier. Infinite symmetric products, function spaces and duality, Ann. of Math., 69 (1959), 142-98.

132. E. Spanier. Algebraic topology, McGraw Hill, New York 1966.

133. N. E. Steenrod. The topology of fibre bundles, Princeton U. P. 1951.

134. N. E. Steenrod (with D. B. A. Epstein). Cohomology operations, Ann. of Math. Study 50, Princeton 1962.

135. N. E. Steenrod and J. H. C. Whitehead. Vector fields on the n-sphere, Proc. Nat. Acad. Sci., 37 (1951), 58-63.

136. E. Stiefel. Richtungsfelder und Fernparallelismus in n-dimensionalen Mannigfaltigkeiten, Comment. Math. Helv., 8 (1935/6), 3-51.

137. J. Strutt. Projective homotopy classes of Stiefel manifolds, Canad. J. of Math., 3 (1972), 465-76.

138. U. Suter. Die nicht-existenz von Schnittflachen komplexes Stiefel mannigfaltigkeiten, Math. Zeitschrift, 113 (1970), 196-204.

139. W. Sutherland. A note on the parallelizability of sphere bundles over spheres, J. London Math. Soc., 39 (1964), 55-62.

140. R. Thom. Espaces fibrés en sphères et carrés de Steenrod, Ann. Sci. Ecole Normale Superieure, (3) 69 (1952), 109-82.

141. P. E. Thomas. On functional cup products and the transgression operator, Archiv. Math., 12 (1961), 435-44.

142. P. E. Thomas. Vector fields on manifolds, Bull. Amer. Math. Soc., 75 (1969), 643-83.

143. H. Toda. Generalized Whitehead products and homotopy groups of spheres, J. Osaka City Univ., 3 (1952), 43-82.

144. H. Toda. Le produit de Whitehead et l'invariant de Hopf, C.R. Acad. Sci. (Paris), 241 (1955), 849-50.

145. H. Toda. Quelques tables des groupes d'homotopie des groupes de Lie, C. R. Acad. Sci. (Paris), 241 (1955), 922-3.

146. H. Toda. On unstable homotopy groups of spheres and classical groups, Proc. Nat. Acad. Sci., 46 (1960), 1102-5.

147. H. Toda. Vector fields on spheres, Bull. Amer. Math. Soc., 67 (1961), 408-12.

148. H. Toda. Composition methods in homotopy groups of spheres, Ann. of Math. Study 59, Princeton 1962.

149. H. Toda. Order of the identity class of a suspension space, Ann. of Math., 78 (1963), 300-25.

150. H. Toda, Y. Saito and I. Yokota. Note on the generator of $\pi_7 SO(n)$, Mem. Coll. Sci. Univ. Kyoto, 30 (1957), 227-30.

151. G. W. Whitehead. On the homotopy groups of spheres and rotation groups, Ann. of Math., 43 (1942), 634-40.

152. G. W. Whitehead. On products in homotopy groups, Ann. of Math., 47 (1946), 460-75.

153. G. W. Whitehead. On families of continuous vector fields on spheres, Ann. of Math., 47 (1946), 779-85.

154. G. W. Whitehead. A generalization of the Hopf invariant, Ann. of Math., 51 (1950), 192-237.

155. G. W. Whitehead. On the Freudenthal theorems, Ann. of Math., 57 (1953), 209-28.

156. G. W. Whitehead. Note on cross-sections in Stiefel manifolds, Comment. Math. Helv., 37 (1963), 239-40.

157. G. W. Whitehead. On mappings into group-like spaces, Comment. Math. Helv., 28 (1954), 320-8.

158. J. H. C. Whitehead. On the groups $\pi_r(V_{n,m})$ and sphere-bundles, Proc. London Math. Soc., 48 (1944), 243-91.

159. J. H. C. Whitehead. Combinatorial homotopy, Bull. Amer. Math. Soc., 55 (1949), 213-45.

160. J. H. C. Whitehead. On certain theorems of G. W. Whitehead, Ann. of Math., 58 (1953), 418-28.

161. L. M. Woodward. Vector fields on spheres and a generalization, Quart. J. Math. Oxford, 24 (1973), 357-66.

162. T. Yasui. On the cohomology of certain quotient manifolds of the real Stiefel manifolds and their applications, J. Sci. Hiroshima Univ., 34 (1970), 313-38.

163. I. Yokota. On the cellular decompositions of unitary groups, J. Inst. Poly. Osaka City Univ., 7 (1956), 39-49.

164. I. Yokota. On the homology of classical Lie groups, J. Inst. Poly. Osaka City Univ., 8 (1957), 93-120.

165. Wu Wen-Tsün. Les i-carrés dans une variété grassmannienne, C. R. Acad. Sci. (Paris), 230 (1950), 918-20.

166. Wu Wen-Tsün. On Pontrjagin classes III, Acta Math. Sinica, 4 (1954), 323-47.

167. P. Zvengrowski. A 3-fold vector product in R^8, Comment. Math. Helv., 40 (1966), 149-52.

168. P. Zvengrowski. Canonical vector fields on spheres, Comment. Math. Helv., 43 (1968), 341-7.

169. P. Zvengrowski. Skew linear vector fields on spheres, J. London Math. Soc., 3 (1971), 625-32.

Index

Adams operations 7, 57, 60, 71-4

Adams conjecture 71

Adem relations 129, 132

admissible vector bundle 57-9

Bernoulli characteristic class 62-8, 73

Bott map 109

Bott suspension 109-15, 121-2

cannibalistic characteristic class 57-9

canonical automorphisms 83-8, 118-19

canonical classes λ, μ 2, 8-9, 18-19, 47, 51, 60-1

canonical line-bundle 36-7, 45, 47, 55, 59, 62-4

cell-structure 21-5

Chern character 62-9, 72-3

Chern class 63-4

Clifford algebra 3, 42-3

Clifford cross-section 3, 18, 81

commutation law 19, 94

complex conjugation 47, 62, 68

composition law 95, 102, 106, 108, 113-14

connectivity 5, 15, 23, 35, 46, 90

coreducible 34

cross-section 2-12, 16, 19, 20, 24, 29, 38-9, 45-8, 53-6, 65, 76, 102-5, 132

decomposable fibration 28-32

derivation law 20, 95-6

differential structure 15

dimension 15

Dold theorem 28, 34, 45-6

duality 49-51, 81-3, 129

EHP sequence 91

evenly connected/related 135-6

exponential characteristic classes 62-70

fibre construction 78-80

fibre suspension 78-81

field of tangent frames/planes 2, 152

generalized EHP sequence 91

generalized Freudenthal theorem 20, 49, 87, 122

generalized J-homomorphism 92

Gram-Schmidt process 15

H-space 151

H^r-condition 146-8

homotopy-commute 123-8

homotopy-equivariance 8, 18, 40-4, 47-8, 76, 130, 133-4

homotopy-normal 124-5

Hopf construction 37, 78-80, 91, 100-4, 130
Hopf line bundle 36-7, 45-7, 55, 59, 62-4
Hopf property 29
Hurwitz-Radon number 3-4, 47
hyperbolic characteristic class 67-70

index of vector field 151
inner product 13
intrinsic join 16-18, 87, 116-22
intrinsic map 17, 49, 51
irregular pair 124-5
iterated suspension 89-90

J-order, etc. 33, 45, 60
J/G-order, etc. 41, 61, 77
$J' = J'_R$-order, etc. 67-76
J'_C-order, etc. 64-70
Jacobi identity 95-7, 110, 112

localization 151

mapping torus 44, 60, 76
McCarty condition 100-1
monodromy exact sequence 44

neutral 9, 40, 133-49, 152

outsimple 133

parallelizability 81, 151
Pontrjagin class 126
Pontrjagin square 125
projective Stiefel manifold 152
pure element (of an algebra) 10-13, 36

quasiprojective space 22

reduced power operation 22
regular pair 124
relative cross-section 53-6, 66
relative intrinsic map/join 54
relative Stiefel manifold 52-6, 66, 109
retractible fibration 27-32, 46, 104-5
retraction theorem 50
row-simple 9

Samelson pairing 123
Samelson product 94-108, 119-24
Serre conjecture 150-1
skew cross-section 8, 18
spin-structure 57
standard fibration 21, 37, 55, 147-9
Steenrod squares 6, 22, 57, 125-32, 135-6, 147-9
Stiefel-Whitney classes 57-60, 125-8

Thom isomorphism 57-8
Thom space 33-7
triad homotopy group 52, 89-90
trivial fibration 8-9, 28-32, 129-32, 154
type (of a map) 129-30, 139, 143-8

Whitehead square 9, 16, 105, 129-31, 136-8, 141-2